On the Origin of Life and Biodiversity

Text, Photos, and Graphics by Mark J. Zamoyski

Drawings by Nicholas J. Lee

ISBN-13: 978-1496036674

ISBN-10: 1496036670

Printed by CreateSpace, An Amazon.com Company

Index / Case File Roadmap

Introduction

In a desert that was once an ocean, 35 fossilized bodies waited for 250 million years to tell their story.

It is the story of the origin of life and biodiversity. This has been one of our greatest planetary mysteries, primarily because of the lack of evidence as to how it occurred. These fossilized bodies provide that evidence.

These fossils provide a new narrative on the origin of life and biodiversity. It is not a story of evolution. The story told by these fossils is of a hosted, random genetic reassortment of unicellular prokaryotic and eukaryotic DNA into biodiverse multicellular life forms. This random DNA reassortment process inextricably links the origin of multicellular life to the origin of biodiversity. The viable life forms spawned by this process effectively serve as the starting point for evolution.

The life forms spawned by this process, as evidenced by these fossils, span the range from unicellular giants to forerunners of dinosaurs and mammals, as well as life forms that never made it into earth's playbook of life.

On an anatomical level, the exceptional level of soft tissue preservation documents the origin of specialized cell types underlying sight and smell, features suited for migration onto land (e.g. reptilian skin, paws, claws) and forerunners of air breathing land animals. Undigested stomach contents reveal that predators are also created as part of this process.

The book takes a detective's approach to reviewing this new evidence and identifies a previously unknown suspect that had the means, motive, and opportunity to perpetrate the explosion of life and biodiversity. The predicted output of this process is consistent with known molecular biology and the fossil evidence.

The story told by these fossils provides a novel perspective on how life on earth arose, our ancestral origins, and how earth repopulated after one mass extinction event with a new cast of characters that did not resemble what lived prior to the extinction event. This novel perspective also has implications for how life may develop on other planets and what it may look like.

The interpretation of the fossils and their molecular biology implications are those of the author alone. Different interpretations and inputs are welcome for consideration.

Chapter 1

The Great Explosion of Life

The Cold Case File:

Suddenly, and seemingly out of nowhere, earth became filled with a large biodiverse collection of multicellular life forms some 500 million years ago, in what is called the "Cambrian Explosion of Life". Explosions of biodiverse life forms have remained one of our greatest planetary mysteries, primarily because of the lack of evidence as to how they happened. Based on new evidence, it is time to reopen the cold case file.

The cold case file is filled with theories about how explosions of biodiverse life forms occur, but because of the lack of hard evidence, there is nothing in the file that would stand up in science court. Just like a crime detective goes after the familiar suspects, evolution is the top suspect in the cold case file, along with variants of evolution in collusion with things like geochemical, environmental, and other processes.

The new evidence points to a new suspect unrelated to evolution. So let's start first by excluding the old suspect. The sudden appearance of new life forms is not something that can be explained by evolution. Simplistically, evolution starts with an existing life form (antecedent lineage) and follows that life form's evolution over time by means of natural selection. It never accounts for how the antecedent life form or lineage arose in the first place. We do.

We need to start with the original antecedent life forms to understand what came first, in order to understand what came next. There were actually three explosions of life on earth. The Cambrian explosion was not the first, it was in fact the last.

A Brief History of Life on Earth:

The Universe is estimated to have originated from the big bang some 13.7 Billion years ago (**Ba**) and the earth formed 4.6 Ba. Three explosions of life occurred in earth's history.

1) The Tough Guys: Prokaryotic cells, or prokaryotes, are bacteria that first appeared on earth 3.8 Ba, shortly after the end of the asteroid impacts.

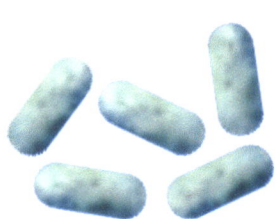 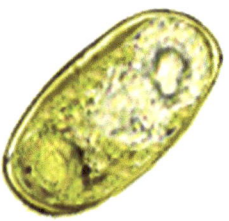 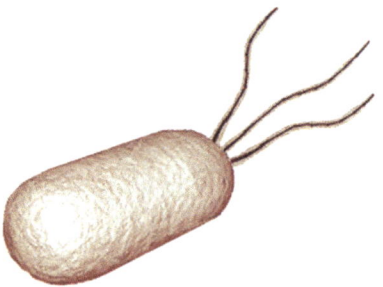

They are single cell organisms, have a tough triple layer cell wall, and can thrive near volcanic vents 3,500 feet below the ocean surface and live two miles deep in soil at pressures of 5,000 PSI. They have circular DNA inside the cell, may have a flagella or whip like tail for propulsion, and can aggregate in colonies that function as a unit.

Cyanobacteria were among the fist to appear and can live in anaerobic or aerobic environments. They obtain their energy by photosynthesis, taking an electron from water and releasing oxygen as a waste product (some can also use hydrogen sulfide). They are believed to have created our oxygen atmosphere.

Cyanobacteria also convert carbon dioxide into carbohydrates, providing a food source, and make the enzyme nitrogenase, which fixes nitrogen into a form that can be absorbed by plants and used in the synthesis of proteins and nucleic acids.

The prokaryotic explosion of life was so successful that today bacterial biomass on earth exceeds that of all plants and animals combined. One gram of soil contains 100 million to 1 billion bacterial cells. Coastal oceans contain 1,000,000 cells per ml (*Li and Dickie, 1996*). A human has 10,000 species of bacteria living in their gut and on their skin and the total number of bacterial cells outnumbers our own cell count.

2) The Soft Guys: Eukaryotic cells or **eukaryotes** appeared 1.5 Ba. They have a soft single layer cell wall and linear DNA that is contained in a membrane bound compartment called the nucleus. Eukaryotes exist as single celled organisms, may have flagella, cilia, or pseudopods for propulsion, and can also aggregate in colonies and function as a unit.

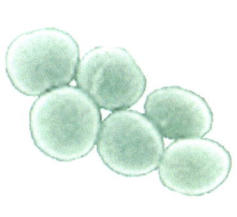 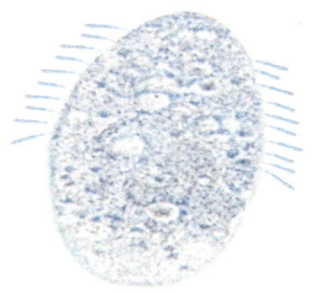 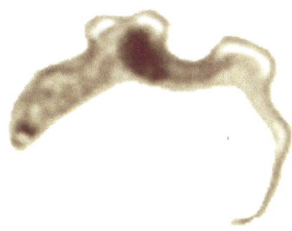

The soft lipid bilayer cell wall is a weak subset of the bacterial (prokaryotic) cell wall. If you push too hard on a eukaryotic cell, it will rupture like a grape. You can hit a prokaryotic cell with a hammer and it will probably still be OK.

Mitochondria is a cell's power plant that stores energy from aerobic respiration (metabolism of glucose). Eukaryotic mitochondria is of prokaryotic origin: its DNA is separate from that of the nucleus, is circular (bacterial), and its nucleotide sequence analysis points back to early bacterial origins (rickettsia, rhizobacteria, and agrobacteria per *Alberts et. al., Molecular biology of the Cell, Third Edition 1994*).

When your favorite detective show talks about matching a suspect's mitochondrial DNA to mitochondrial DNA found at the crime scene, what they are really saying is "Which combination of bacteria does our suspect hail from, and does that combination match the one found at the crime scene?"

Since eukaryotic cells took mitochondria and mitochondrial DNA from bacteria, have a cell wall that is a subset of the bacterial cell wall, and exhibit many of the same features and traits (which are the result of DNA expression) as their prokaryotic counterparts, the reasonable inference from the evidence is that eukaryotic cells hail, in whole or in part, from prokaryotic origins.

3) The Collective Unit: Multi Cellular Life Forms are a collection of different eukaryotic cell types comprising a single life form (plants, worms, fish, animals, humans, etc...). Each cell contains the complete DNA code for all of the cell types, but expresses only its subset of that DNA. A human is a collection of ~210 different eukaryotic cell types.

Multi Cellular Life first appeared with a vengeance around 500 million years ago (**Ma**) during a biodiversity explosion event known as the "Cambrian Explosion of Life".

Biodiversity explosions also appear to happen after extinction events. During the Permian period, the earth experienced 4 extinction pulses, each of which was followed by a biodiversity recovery pulse *(Sahney and Benton, 2007, summarized in the graph below for terrestrial tetrapods)*.

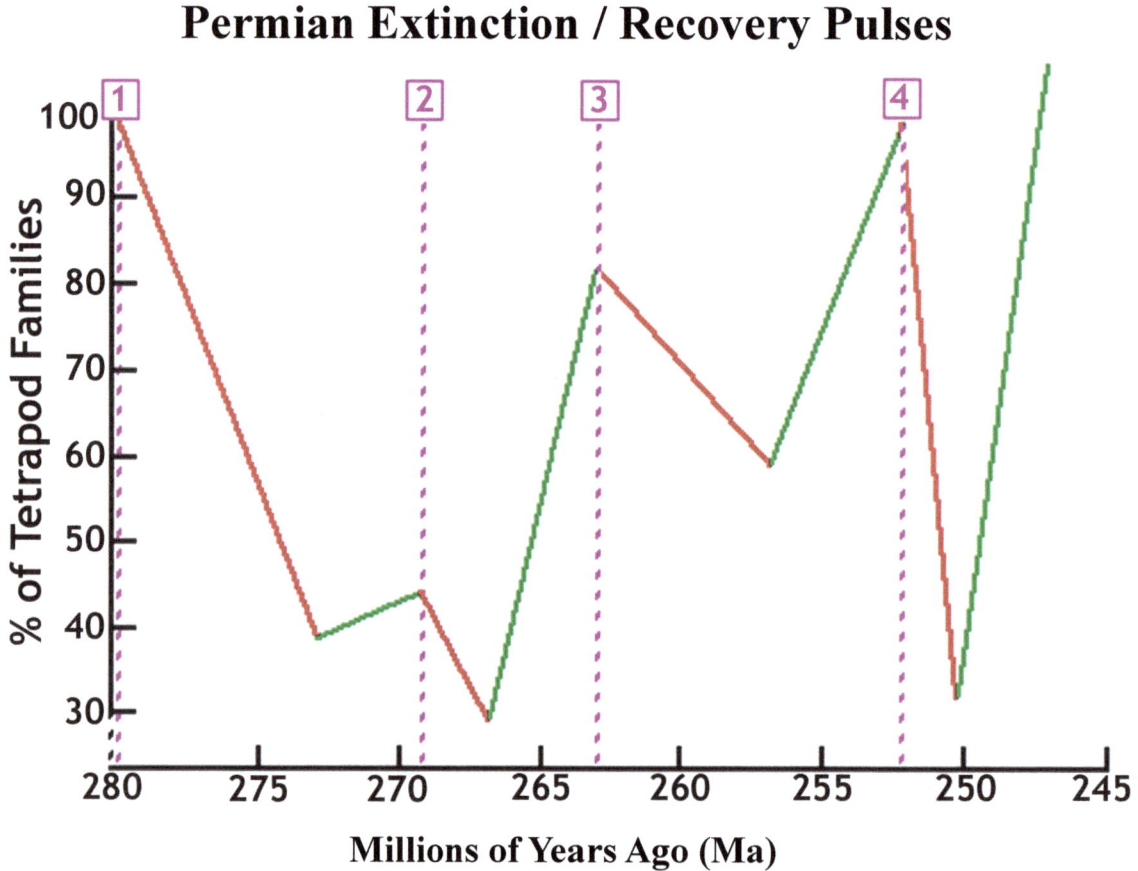

The last Permian extinction (Permian - Triassic or P-Tr) event 252 Ma was the worst and it was worse for marine life than for terrestrial life. Despite the 95% marine extinction rate, marine life re-speciation to pre-extinction levels occurred in as little as 2 million years *(ScienceDaily, Oct. 5, 2011)*.

Such a rapid appearance of new, biodiverse life forms, without antecedent lineage, is something evolution cannot do. This brings us back to clearing the old suspect from the file and identifying a new suspect.

In order to develop a profile of the suspect that could have spawned such biodiverse multicellular life, it is first necessary to understand DNA and DNA expression. DNA and DNA expression determines what a life form is and what it does.

Those familiar with DNA can skip the next paragraph. Everyone else, hang on to your seats as we are going to take a bullet train ride through a molecular biology book.

DNA and DNA Expression

DNA is the blueprint for proteins. **DNA expression** means synthesis of proteins from that blueprint. Synthesis of proteins in eukaryotic cells is achieved by a process that involves 1) Transcription of DNA into mRNA in the nucleus, 2) Transport of the mRNA strand to the ribosome (made up mostly of rRNA), and 3) complimentary base pair binding of tRNA with an attached amino acid, whereby the mRNA strand is translated into a protein. Proteins make up 60% of a cells dry mass and determine what a cell does.

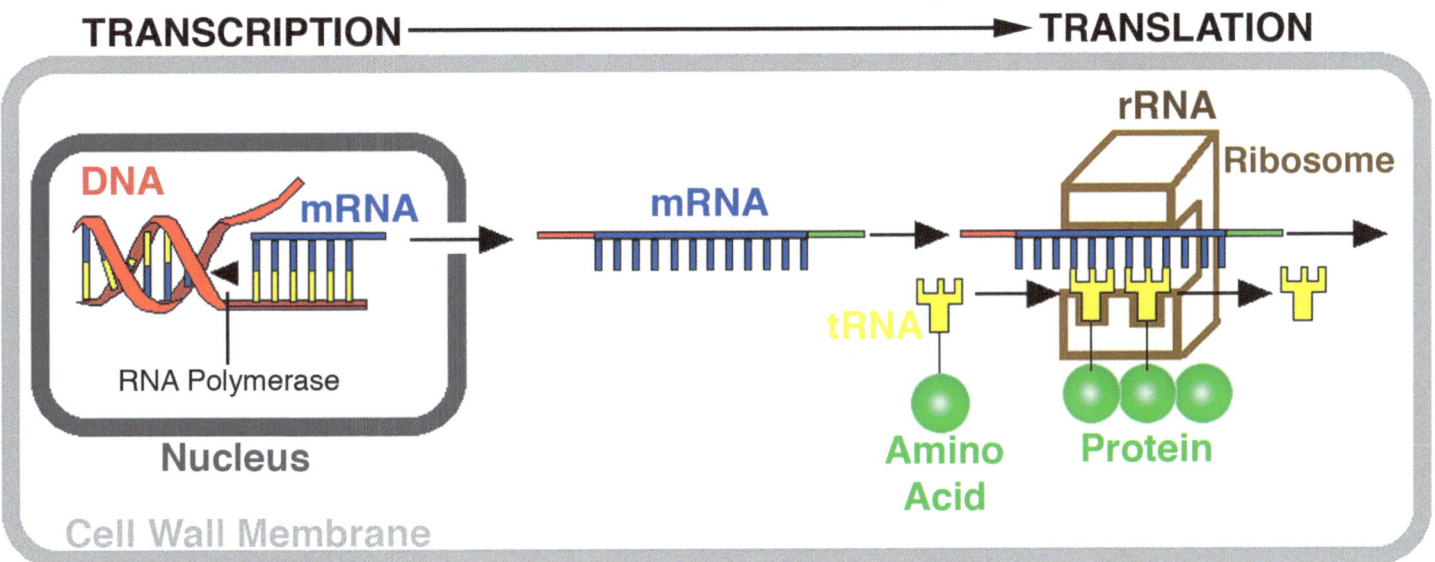

Genomic Efficiency

A simple measure of genomic efficiency can be made by comparing how many proteins are synthesized per million base pairs of DNA.

The cells with the best genomic efficiency could be argued to be the most advanced.

So how does today's linear eukaryotic DNA compare to the 3.8 billion year old circular prokaryotic DNA?

The genomic efficiency of ancient prokaryotic cells *(see Appendix A)* versus a human eukaryotic cell *(DOE, Human Genome Project, Oct. 2004 findings)* is summarized below.

	DNA Base Pairs	Known Proteins	Proteins Per Million Base Pairs
HUMAN	3,000,000,000	21,787	7
Cyanobacteria	5,872,274	5,238	**898**
Rickettsia	1,295,744	1,163	**896**
Burkholderia	6,608,450	5,851	**896**
Archaea	1,925,255	2,113	**1,098**

Well isn't that interesting.

The supposedly superior human eukaryotic cell cranks out only 7 proteins per million DNA base pairs versus bacteria that crank out around 900 proteins per million base pairs. The 3.8 billion year old cyanobacteria's circular DNA is some 130 times more efficient than the linear human DNA. Archaea, the oldest known prokaryotic cell, is 156 times more efficient.

The human genome project also revealed that 98% of human DNA is non-coding (i.e. not used). We are basically a genetic junkyard, with a few good sequences.

The molecular biology previously presented indicates eukaryotic cells hail from prokaryotic origins. The progression from unicellular to multicellular eukaryotic life, in turn, could only have started at a unicellular level. Given that multicellular life requires a radically reassorted and inefficient genome per above, the probability of creating a single <u>viable</u> eukaryotic cell that coded for a multicellular life form would be small. To expect this exact DNA reassortment to happen simultaneously in every cell type and every cell of a multicellular life form is beyond what is mathematically possible.

Suspect Profile and M.O.

We can now start constructing a suspect profile. Who or what could have done this?

The suspect we are looking for must be capable of committing horrible atrocities against DNA, starting at a unicellular life form level. Our suspect's M.O. would likely be to arbitrarily shred and recombine DNA, creating Frankenstein cells in a "roll of the dice" fashion, with a large percentage of the DNA not being usable. We will call this crime "unlicensed DNA reassortment". Most of the cells created by this unlicensed DNA reassortment process would never live to tell the story of what was done to them. A rare few would be viable, survive, and go on to become the genetic junkyard underlying life as we know it today.

The breaking news is that a new "unlicensed DNA reassortment" scene has just been found, with lots of victims and lots of evidence this time, so that's where we are off to in the next chapter.

Chapter 2

The New Hard Evidence

The basic investigative questions must be answered first.

Where did the event happen?
The event happened in Northern Arizona. Thirty five bodies have been recovered from the scene.

What is the event?
A preliminary review reveals an event on top of an event. An unlicensed DNA reassortment event was in progress when the mass murder of everybody occurred.

The scattered bodies include both victims and perpetrators. Tools and weapons used to commit both the unlicensed DNA reassortment and the murder of everybody are also present.

We will pursue only the unlicensed DNA reassortment case, leaving the mass murder for another day.

When did the crime occur?
The bodies are marine life forms. The site was intermittently a shallow ocean from around 550 to 250 Ma *(Blakey and Ranney, 2008)* before tectonic activity lifted the state to form dry land.

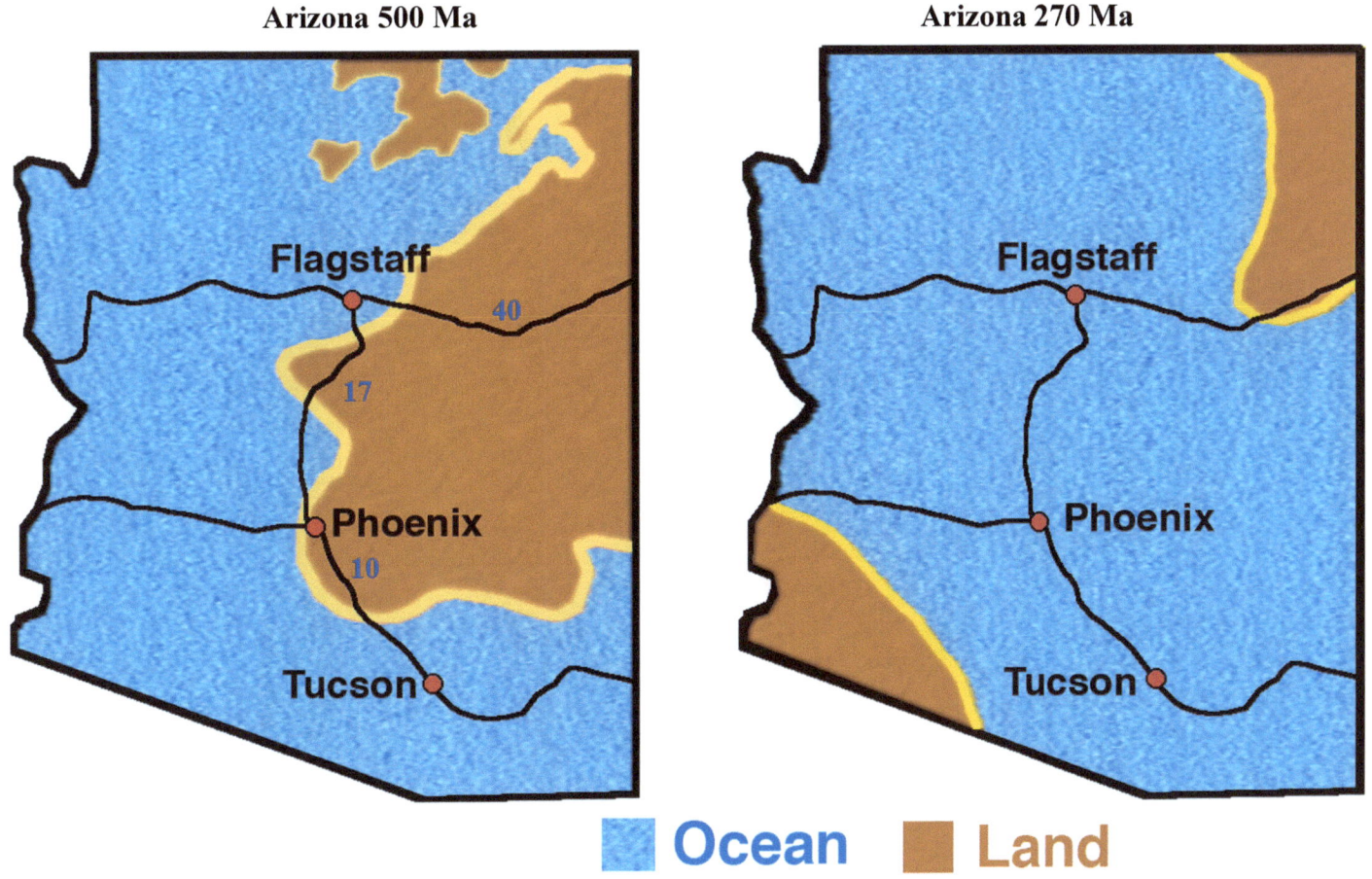

Tectonic activity, combined with erosion, left the ancient ocean floor at the surface in parts of Arizona.

The U.S. Geological Survey Time and Terrain maps date the age of surface soil at the site as Permian, putting the time of the event between ~250 and 280 Ma.

This tells us the event happened during the Permian extinction/ recovery pulses.

Age of Surface Soil Today

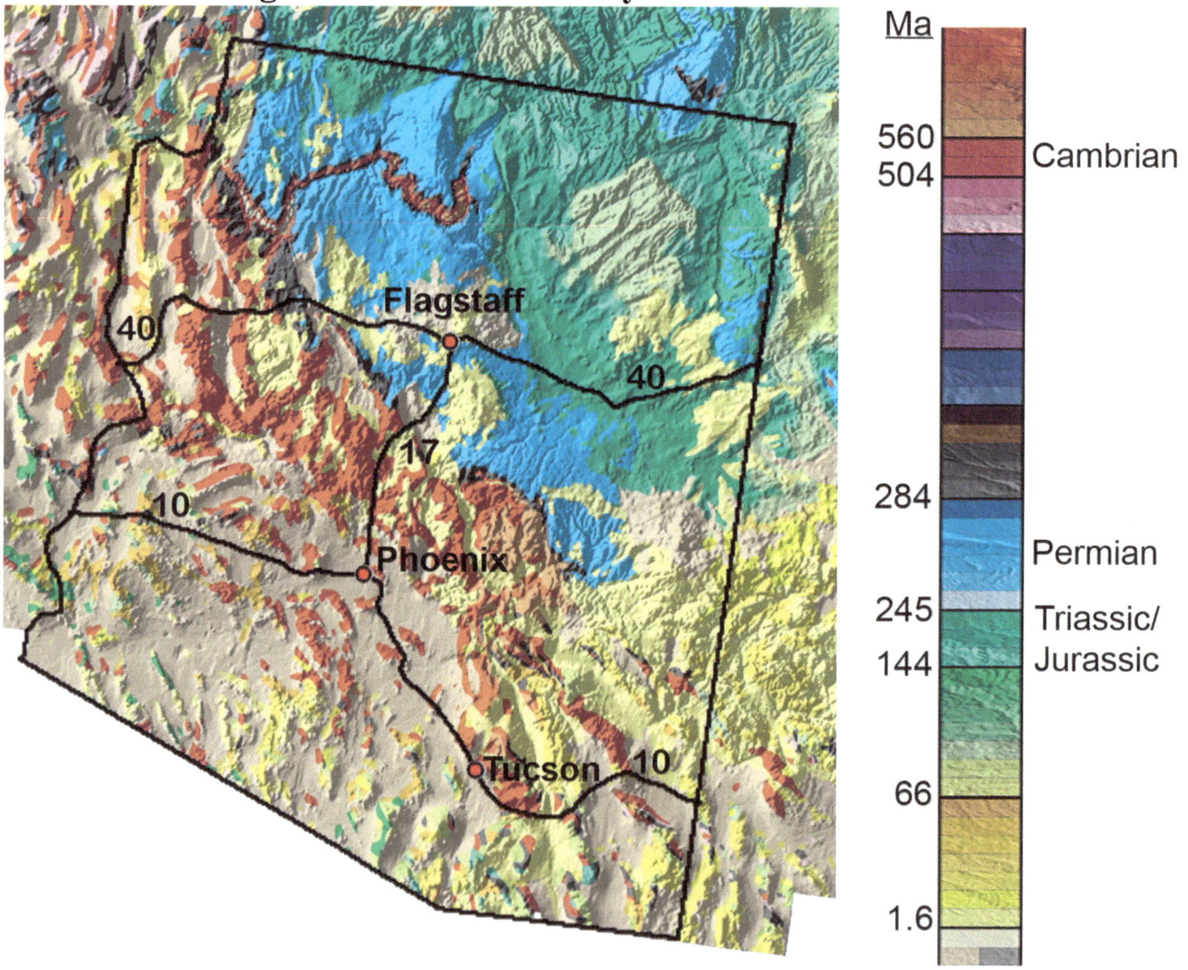

Now the detailed crime scene investigation can begin. We start by sorting through all of the bodies, to see if we can separate the victims from the perpetrators. The bodies have become entombed in calcium over the last 250 million years. From the outside, only the tomb is visible. An example is shown below.

We cut the tomb into halves to reveal the life form inside. This "sectioned specimen" is shown below:

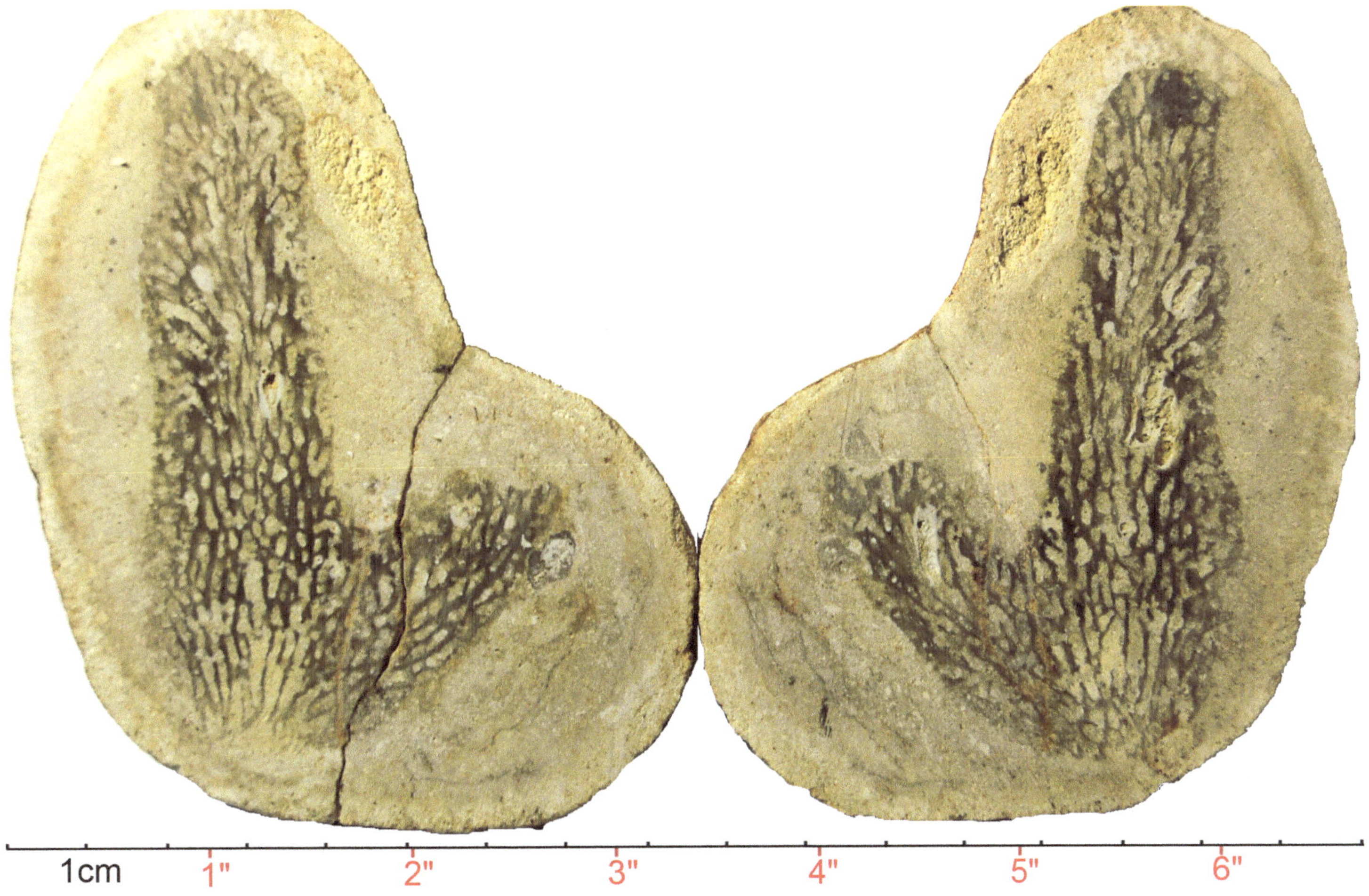

1cm   1"   2"   3"   4"   5"   6"

The lighter parts are the encasing calcium matrix or tomb. The darker parts are the organism. Silicon has replaced the calcium in the organism.

As it turns out, the suspect above is known to us.

This suspect is morphologically identical to the Cambrian Protopharetra, an irregular archaeocyathan, *(Boardman et. al., Fossil Invertebrates, 1987, p. 114 Fig. A and B),* with the notable exception that this suspect is about 5 times larger than its Cambrian counterpart. Archaeocyathans were the first unicellular eukaryotes that lived in colonies and secreted calcium carbonate as skeletal material. They were filter feeders, channeling moving ocean water through the hard skeleton labyrinth.

Our suspect's counterpart first appeared at the onset of the Cambrian biodiversity explosion event and then disappeared under suspicious circumstances as the new ecosystem emerged. Our suspect's reappearance during the Permian biodiversity explosion event may be a coincidence. Skeptical of coincidences, we will name this suspect "**Permian Protopharetra**", and put it on our suspect list for further investigation.

As part of the crime scene investigation, we also use a forensic artist to develop drawings from the remains to help visualize what they likely looked like when they were alive.

When external features are well preserved the illustrations are likely accurate. In instances where illustrations are constructed only from internal features, there is more room for error. The true color of the life forms cannot be determined with certainty, as atoms that give color may have been lost during the fossilization process and other atoms added during the fossilization process. The color is in large part the artist's choice.

As an example, the artist's rendering of the **Permian Protopharetra,** constructed from the above sectioned specimen photo, is shown below:

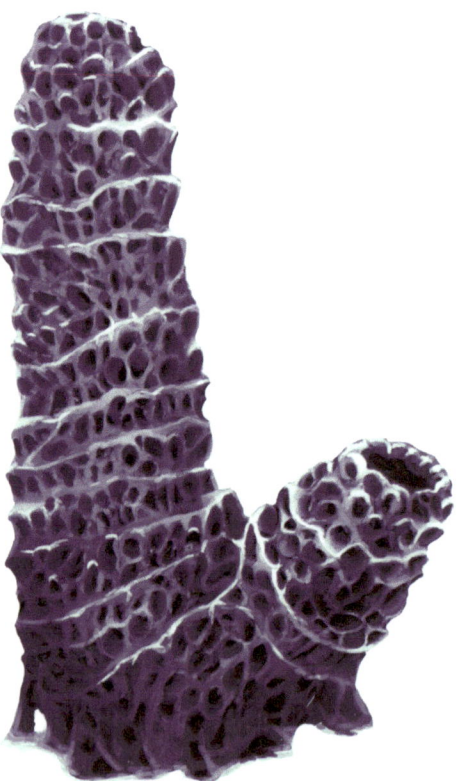

We have given names to each of the bodies found. We have put the drawings of all of the bodies up on a "Working Wall" in our office as we start sorting through this mass murder.

Murder Scene Working Wall: The Bodies

Dividing Uni-cellular Giant

Flagellate Uni-cellular Giant

Big Blob

Sea Pig

Sea Piglet

Central Question

Tongued Red

Blue Jelly

Mud Worm

Braniac Worm

Sea Snake

Jawless Fish

Double Fish

Flagella Fish

Goby Shark

Whaleen

Claw-Paw Dragon

Zamoyski Dragon

Geckosaurus

Bird-Squito

Arizona Lung Fish

Dino-Seal

Aqua Duck

Eel-a-Phant

Sea Horse

Permian Protopharetra

Spherical CSFF

Mega Protoparetra

Columnar CSFF

Carnivorous Columnar CSFF

Large Encased CSFF

Internal Soft Tissue CSFF

Carnivorous Protopharetra

Protruding CSFF

Encased Spherical CSFF

The Y

The bodies have been sectioned and autopsied. It is now time to review the autopsy results of our victims, to see what was done to them, in order to provide insights into our DNA reassortment perpetrator's M.O.

A pattern begins to emerge. DNA capable of coding for unicellular giants and life forms composed of two or three cell types begins to appear. Based on the timeline previously established, we can place our victims at the scene during the Permian extinction / recovery pulses. The reasonable inference is that the unlicensed DNA reassortment may have been driving one of the biodiversity recovery pulses. The mass murder may have been the result of a subsequent extinction pulse.

The presentation of the life forms with reassorted DNA roughly follows increasing DNA complexity, as evidenced by the features resulting from the expression of that underlying DNA. These features include motility, vision, smell, skin, bone, reproduction, pre-Triassic / Jurassic features, and pre-terrestrial features.

Motility / Propulsion

Motility (ability to move) is a trait common to both prokaryotic and eukaryotic cells. Unicellular organisms use flagella, cilia, or pseudopods for motility. Eukaryotic flagella and cilia are structurally identical.

The human sperm cell uses a flagellum to propel itself. In humans, cilia drive the movement of the mucus blanket that sweeps dirt out of the lungs. Beating of the cilia in the fallopian tubes moves the egg from the ovary to the uterus.

The rigid three layer cell wall of prokaryotic cells could only accommodate a rear mount flagellum for motility:

Bacteria

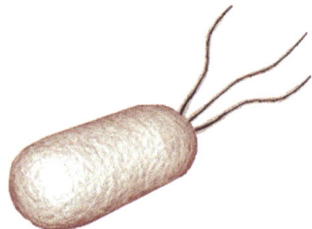

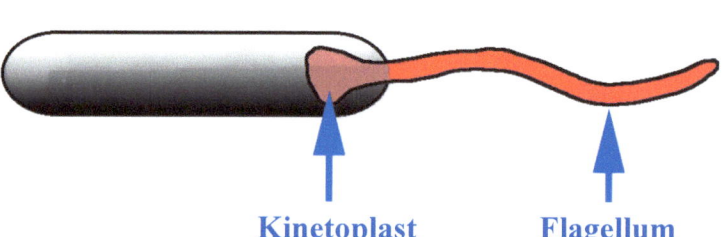

Kinetoplast   **Flagellum**

The flexible single lipid bilayer cell wall of eukaryotic cells allowed for either a rear mount flagellum or a side mount flagellum. A side mount flagellum is shown below in the unicellular trypanosoma (which causes sleeping sickness):

Trypanosoma (~ 25 μm long)

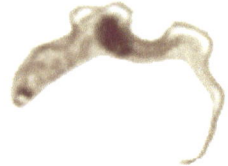

Nucleus   **Kinetoplast**   **Flagellum**

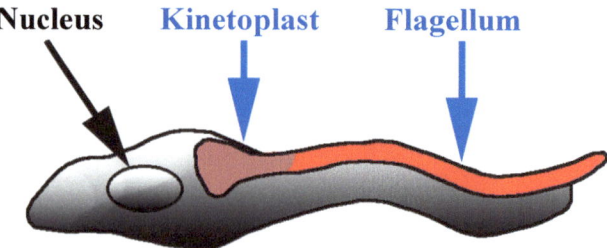

Flagellate Unicellular Giant is an example of a unicellular specimen with a rear mount flagellum. At 2" (5 cm) long, it is ~ 2,000 times larger than the unicellular trypanosoma above. Unicellular giantism may have arisen from altered DNA expression related to growth control pathways or cell division pathways.

The author's photo derived rendering of what the **Flagellate Unicellular Giant** looked like:

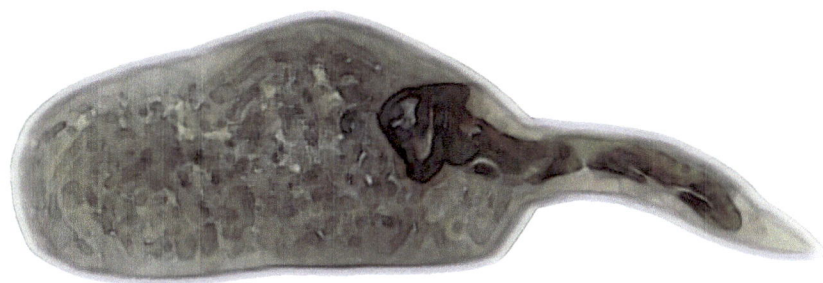

An enlarged photo of the sectioned specimen, in matrix (stone), is shown below:

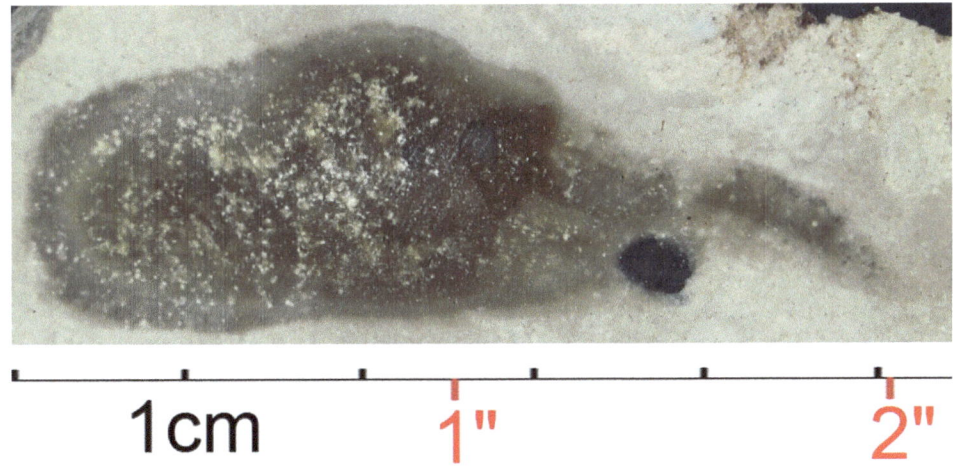

The above photo, with the matrix excluded and rear mount flagellum highlighted, is presented for clarity:

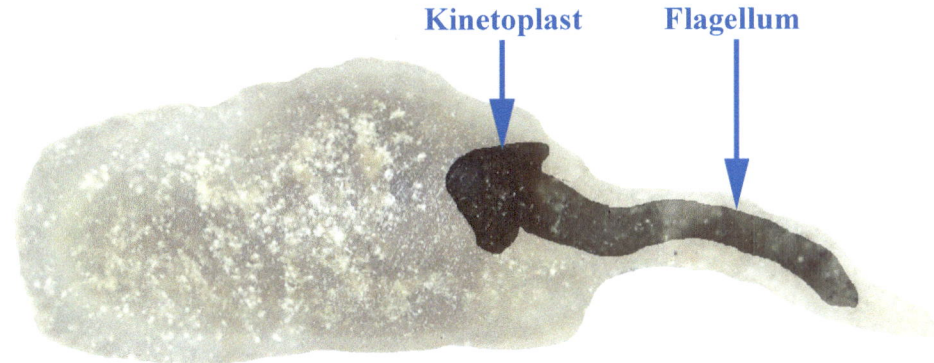

Flagella Fish is an example of a side mount flagellum in a multicellular organism. Flagella Fish is ~5,000 times larger than the unicellular trypanosoma.

The artist's rendering of Flagella fish is provided below:

Flagella Fish Illustration

A photo of the actual sectioned specimen is shown below. **Flagella Fish** appears to have a cartilaginous body with arm like protrusions at the front and a side mount flagellum for a tail.

Sectioned Specimen Photo - Matrix Digitally Excluded

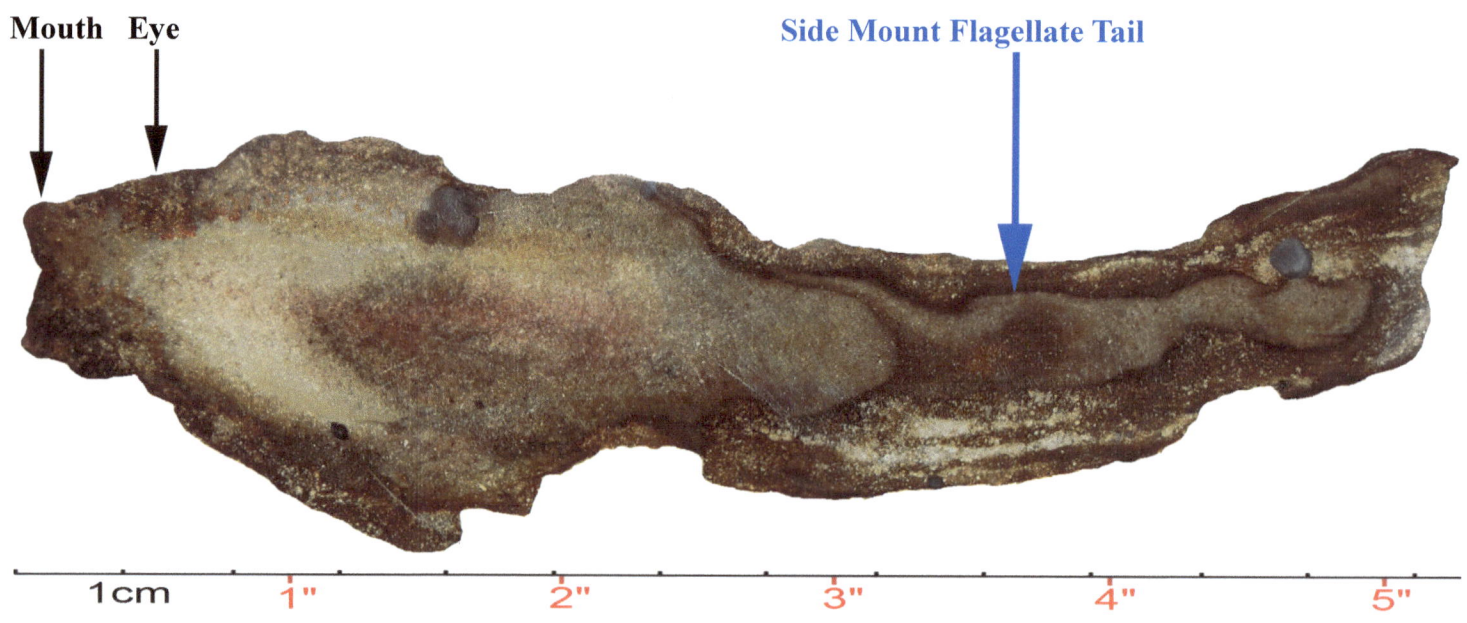

14

A zoom of the tail shows the flagellum integrated into the side of the body.

Side Mount   **Flagellum**

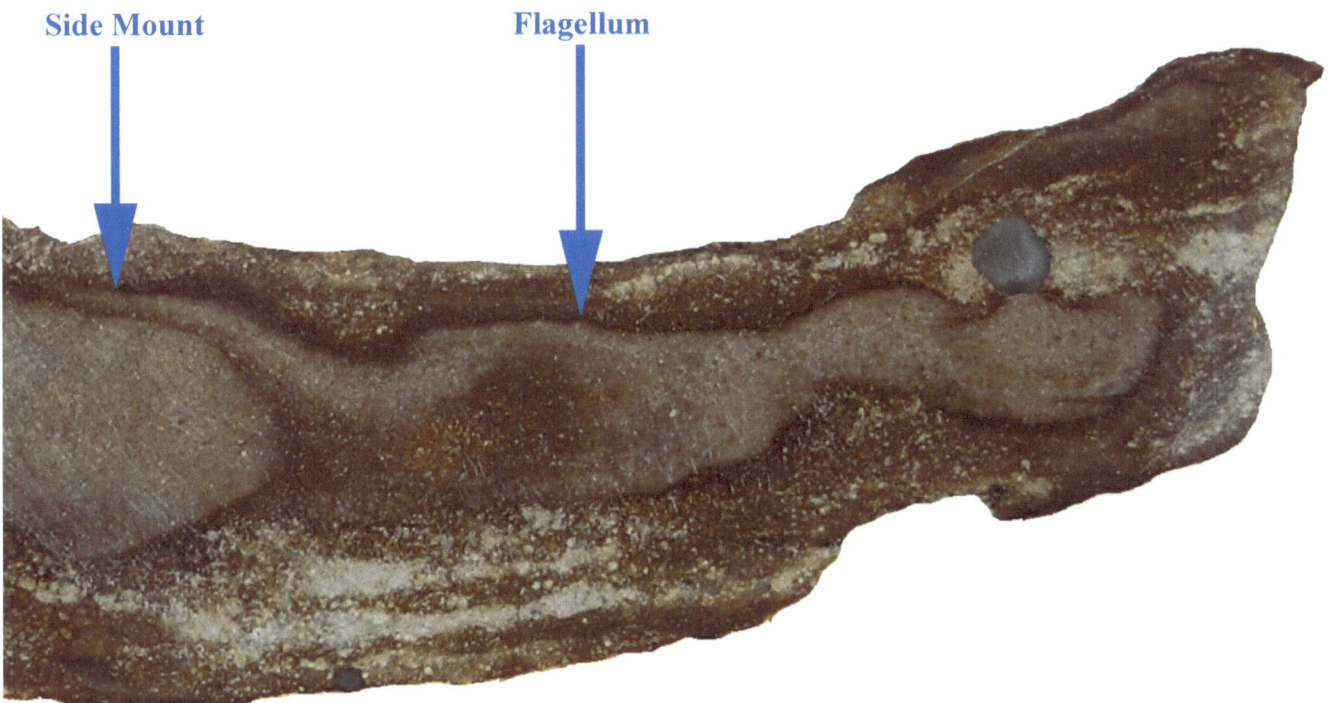

A zoom of the body shows the arm like protrusions, with a possible paw / paddle at the end.

Mouth   **Eye**

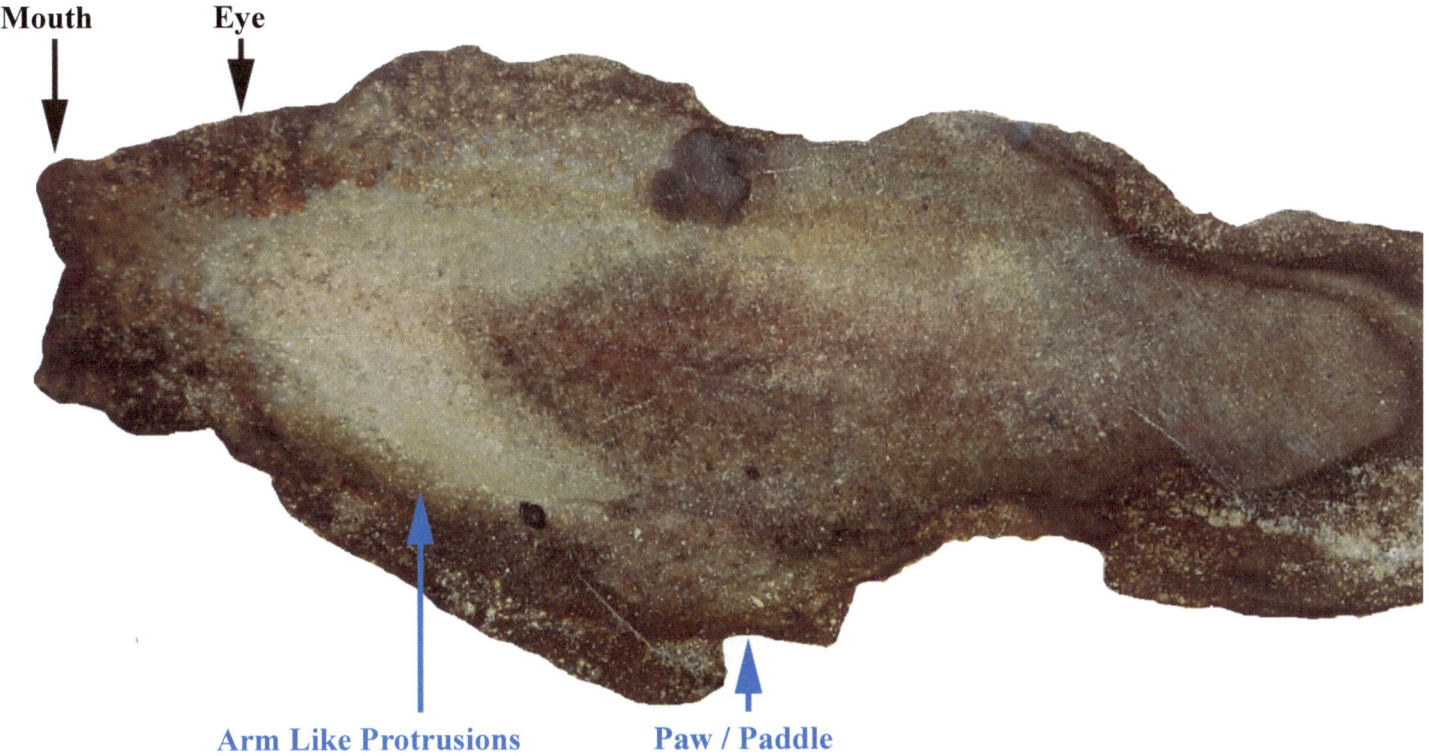

Arm Like Protrusions   **Paw / Paddle**

A cyclops like eye at the top left would likely indicate Flagella Fish had a phototactic guidance system. Specimens with more visible optic neurons and optic structures are covered shortly.

Blue Jelly is an example of another propulsion system. Jellyfish are not truly fish, but are one of the simplest multicellular life forms. They are a pulsating gelatinous bell with long trailing tentacles.

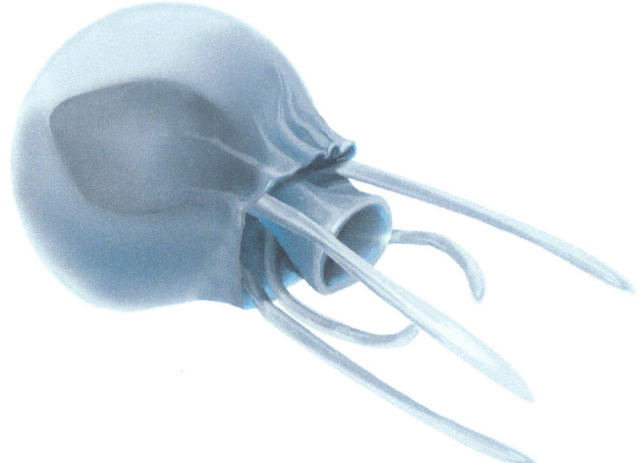

A photo of the actual sectioned specimen in matrix (~ 3" long) is shown below. The mushroom like central structure appears to be in the contracted phase (propulsion) and is isolated for clarity below.

Sectioned Specimen in Matrix

Isolated Propulsion System

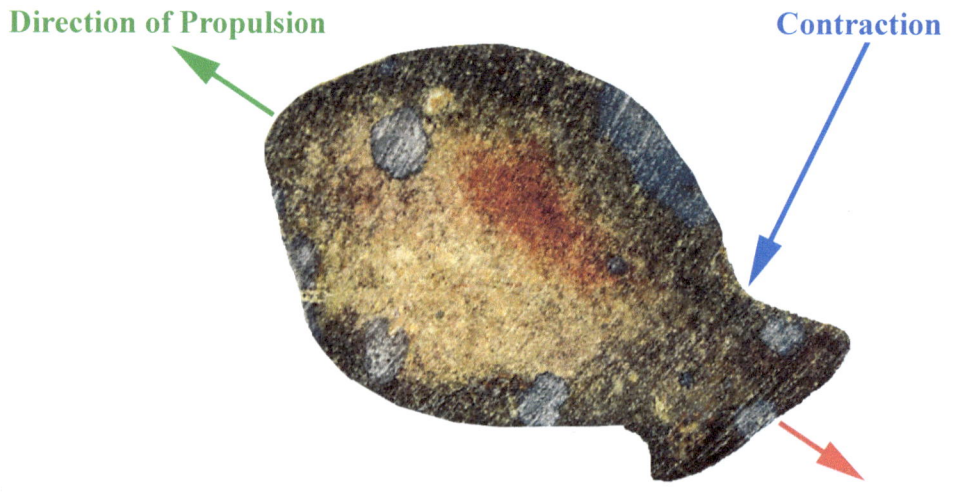

Direction of Propulsion

Contraction

Ejected Jet of Water

Phototaxis / Vision

Phototaxis is the initiation of motility (activation of flagella, cilia, or pseudopods) in response to light. Chemotaxis is the initiation of motility in response to chemical signals.

Cells that use photosynthesis have the ability to orient themselves toward light. That trait is also observed in non photosynthesizing cells. In the ocean, it is advantageous to move towards light, as that is where the most abundant food is (e.g. where photosynthetic organisms / algae thrive).

A specialized optic neuron cell begins to appear in these early life forms. A neuron is a specialized cell that receives, conducts, and transmits signals.

Neurons maintain an electrochemical gradient between the inside of the cell and the outside of the cell. In a sensory neuron, the dendrites gather sensory information. If a threshold level is reached, a traveling electrochemical perturbation is initiated along the axon. When it reaches the synapses, voltage gated channels transiently open allowing an inrush of calcium ions that cause a neurotransmitter to be released. The electrical signal is thus converted into a chemical signal. If the neurotransmitter is acetylcholine and the terminal branches / synapses integrate into muscle fiber, a muscle contraction would result.

Neuron   **Zoom of Synapse**

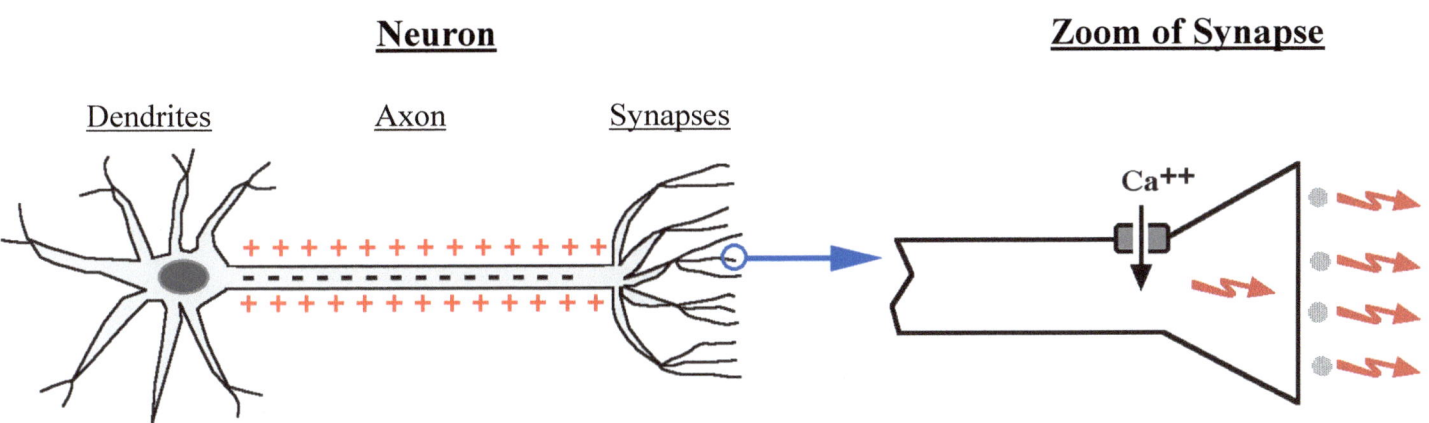

Dendrites   Axon   Synapses

An example of an isolated optic neuron found in one of the fossils is shown below:

Eye   Axon   Synapses

The specimen that displays this optic neuron is called **Whaleen**. Whaleen resembles a truncated Baleen Whale. The artist's rendering of Whaleen is provided below:

The photo of the actual specimen is shown below. The specimen is ~ 5" or 12 cm long.

Exterior View - Side A

Eye / Phototaxis

The optic neuron can be seen from the inside view of the sectioned specimen:

Interior View - Zoom of Side B Optic Neuron

The optic neuron is further isolated in the photo below:

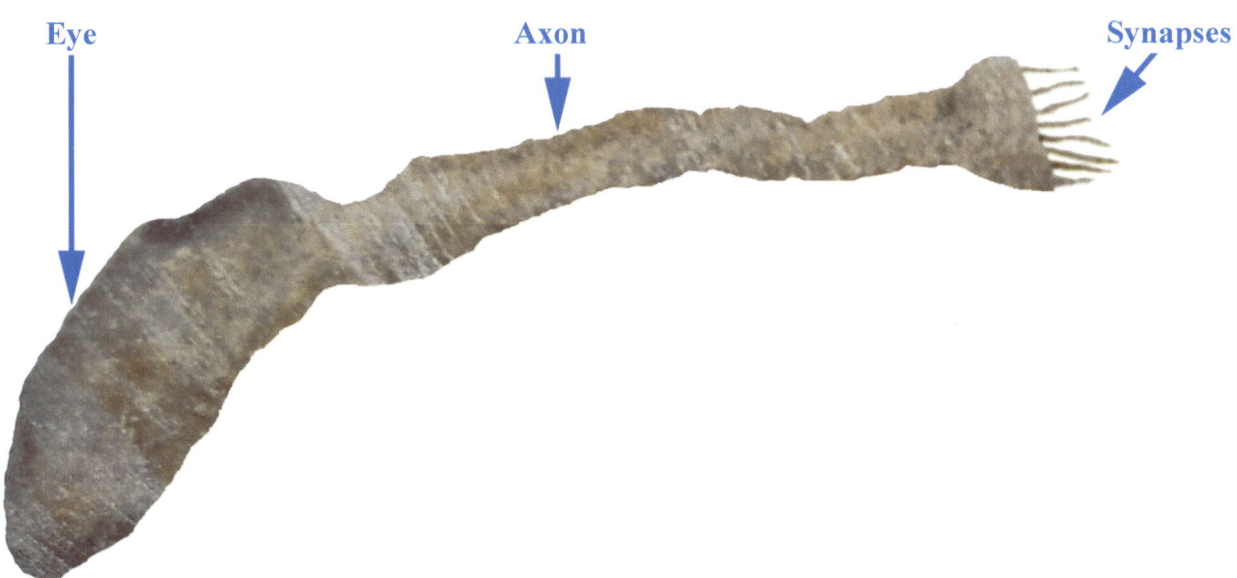

This early neuron likely conveyed light information to the tail to propel the organism towards light. In water, light is where the food is. This primitive response of moving toward light may be what drives insects to your porch light at night.

The above phototactic neuron may in turn have been an adaptation of an earlier phototactic organelle, which looks like a truncated neuron, and is seen in a specimen called Sea Pig , shown below, and included in the chapter or Unicellular Giants and Indeterminate Life Forms.

Sea Pig Illustration

Sea Pig Photo - Interior View, Matrix Digitally Excluded

Isolated Phototactic Organelle

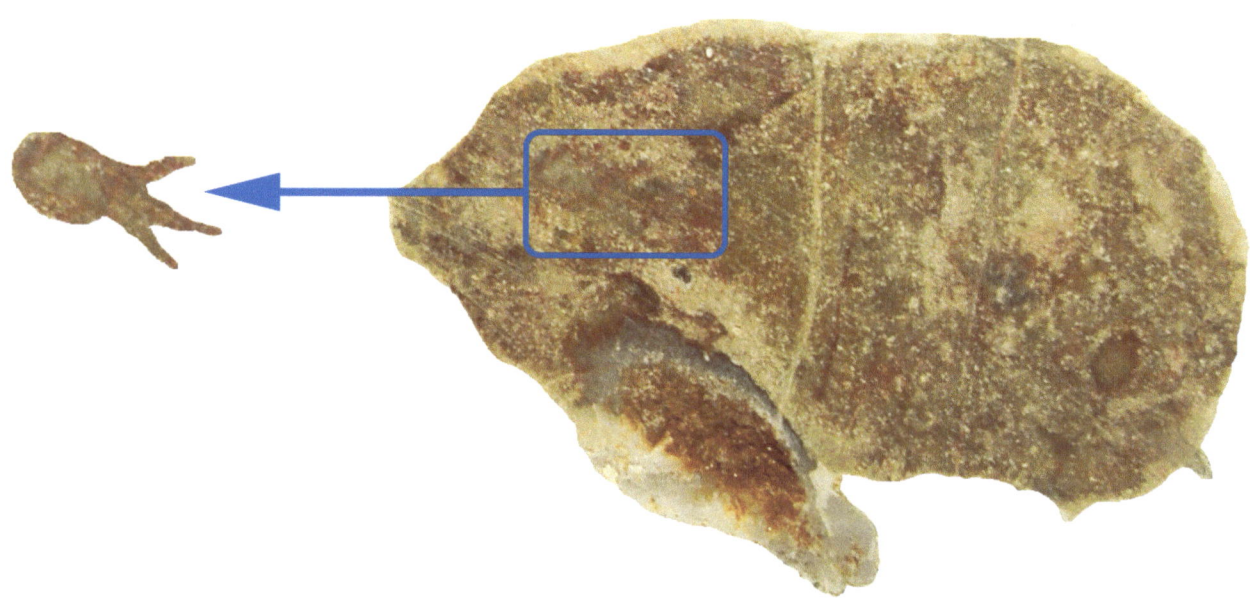

Chemotaxis / Smell

Chemotaxis is the initiation of motility (activation of flagella, cilia, or pseudopods) in response to chemicals in the environment. This is used by both prokaryotic and eukaryotic cells to find and move toward food and to flee from poisons (negative chemotaxis). As an example, amoeba feed on other protists, algae, and bacteria and exhibit chemotactic responses to glucose and cAMP.

In humans, cells such as neutrophils, the body's first line of defense against bacteria, recognize chemicals produced by bacteria and move directly toward them. Additionally, tissue resident mast cells are activated by antigens and in response release chemotactic factors such as Eosinophil Chemotactic Factor A, Chemotactic Factor (NCF IL8) and Leukotriene B4, which in turn result in chemotaxis of a broader set of immune system cells toward the site.

At a multicellular organism level, chemotaxis is performed by neurons.

Dino-Seal appears to have such an olfactory (chemotaxis) neuron, in addition to a phototactic neuron. An artist rendering of Dino-Seal is provided below:

The actual specimen is ~ 4" or 10 cm long, and the vertical undulation orientation of the tail is characteristic of all marine mammals (e.g. sea lions, seals, dolphins, whales), suggesting Dino-Seal is a mammal forerunner.

The exterior and interior photos of the actual specimen reveal an increased level of multicellular complexity, as shown next:

Sectioned Side A: External View

Nose   Eye   Tail

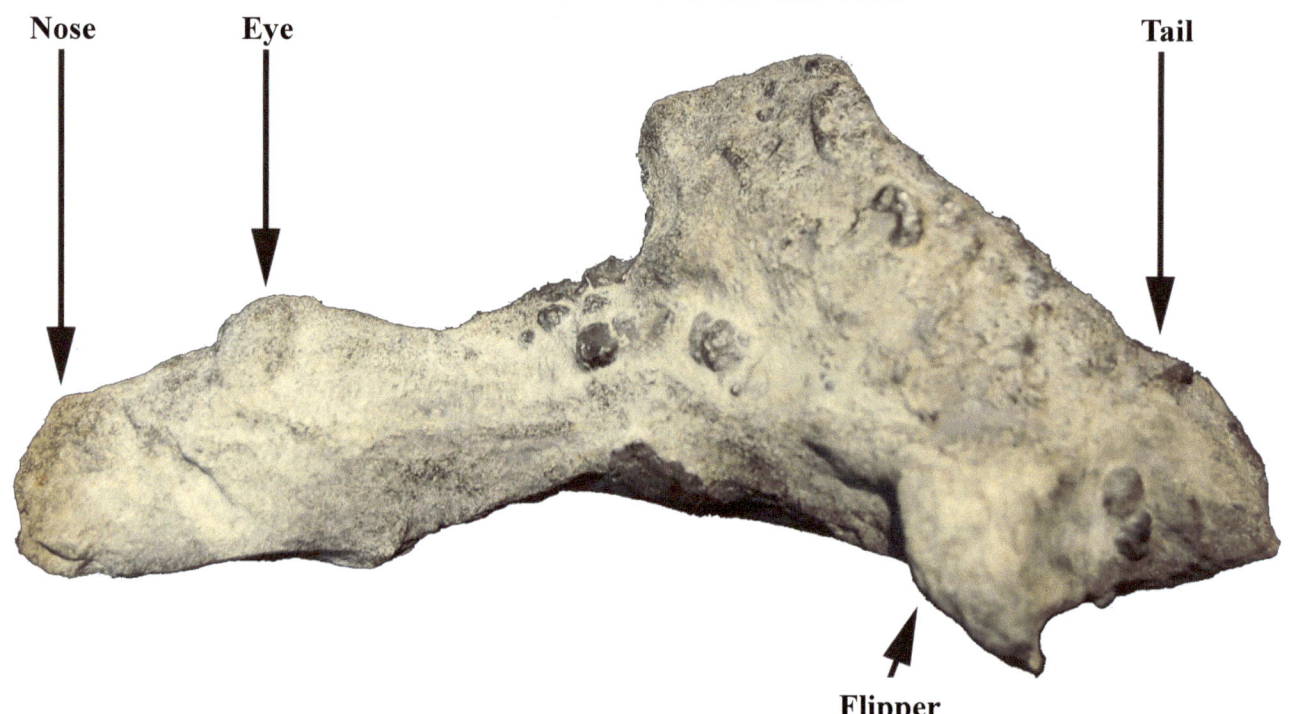

Flipper

Sectioned Side B: Interior View, Matrix Digitally Excluded

Phototaxis   Axon   Tail

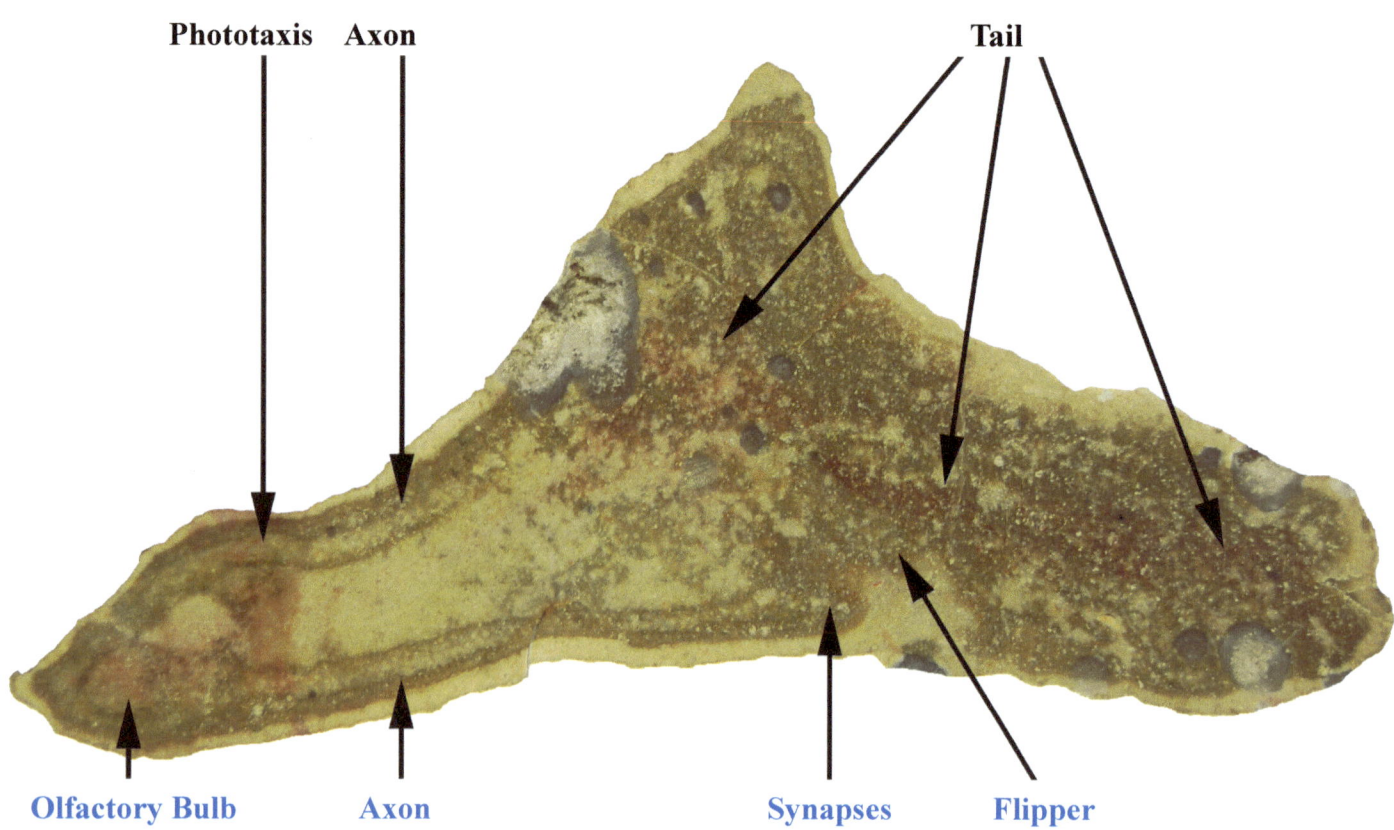

Olfactory Bulb   Axon   Synapses   Flipper

Dino-Seal appears to have a phototactic neuron at the top, but a large crystal formation has obliterated the synapse portion of the neuron. A chemotactic neuron starts in the nose / mouth area and runs to the flipper, likely activating motion of the flipper in response to smell so it would swim toward what smelled good.

A zoom with the olfactory neuron highlighted and isolated is shown below:

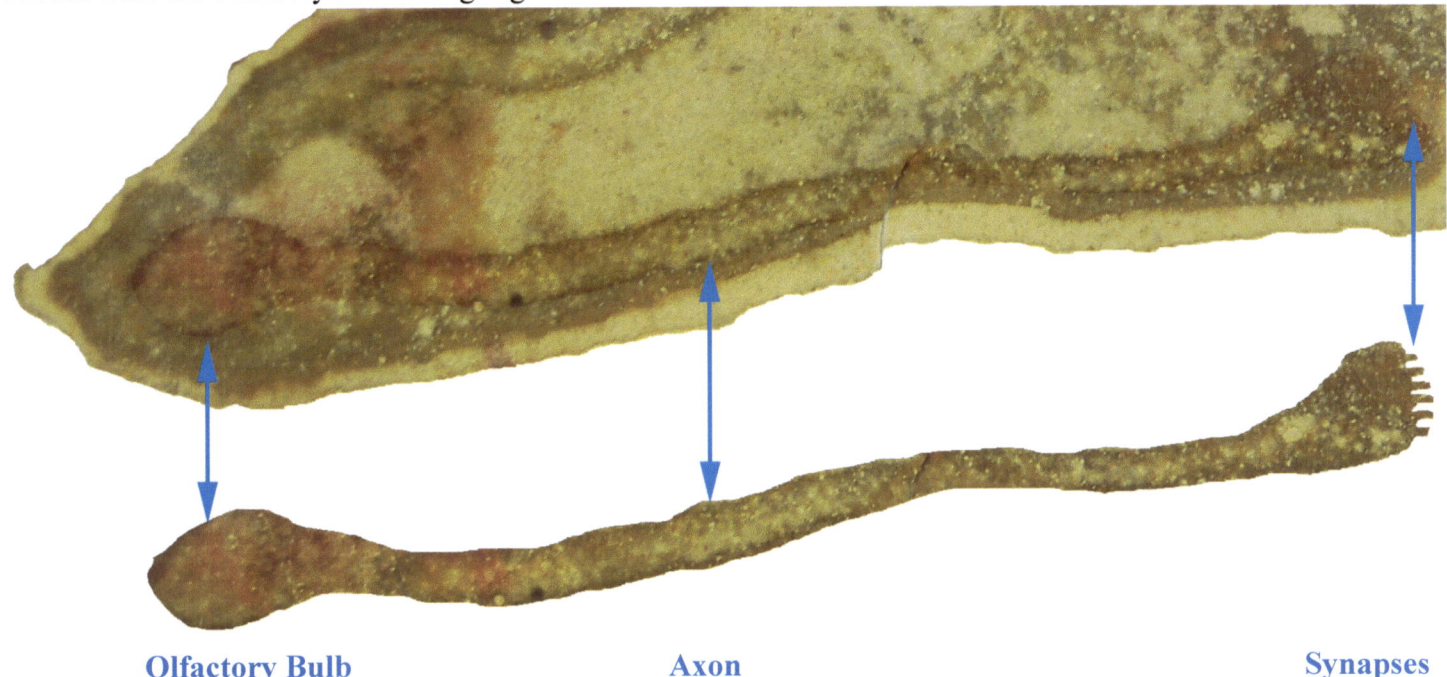

Olfactory Bulb   **Axon**   **Synapses**

Skin

Unicellular organisms have cell signaling capability in colony situations that causes certain cells to differentiate and form a hardened protective outer layer.

Skin is an important feature for multicellular life. A 70 kg human (~ 70 liters volume) is made up of ~ 10 liters of cells (~ 10 trillion cells) bathed in 40 liters of extracellular fluid, with the balance made up of bone, fat, muscle fibers and connective tissue. Everything is contained within a "skin sack". The resident extracellular saline solution is our gulp of the ocean we needed to take before we could step onto land.

An aqueous environment allows atoms to exist as ions (Na^+, Cl^-, K^+, Ca^{++}) which in turn allows for maintenance of concentration gradients including electrochemical gradients. On dry land, atoms such as Na and Cl combine to form electrically neutral NaCl, or table salt.

On an atomic level, physiological life can be defined as a collection of concentration gradients. The presence of concentration gradients means life. The absence of concentration gradients means death.

Skin also revolutionized cell signaling. Cell signaling is the production of chemicals (e.g. testosterone, estrogen) by a cell that alters DNA expression of distant cells. In a closed environment, the chemical signals are not washed away by the ocean, but can more effectively reach their intended target cells.

In humans, the epidermis is a columnar stack of cells, with roughly one cell a day shedding off the top of the column, and a specialized pluripotent cell at the base dividing daily to maintain the stack depth. Below the epidermis is the dermis, which is mostly connective tissue, produced by a cell type known as a fibroblast.

A harder skin allows for survival in harsher or more abrasive environments, such as mud or land. An example of a hardened protective exoskeleton can be seen in the **Zamoyski Dragon**, which is ~ 7" (18 cm) long.

The artist's rendering of the **Zamoyski Dragon** is provided below:

The stomach contents of the **Zamoyski Dragon** reveal an undigested fish inside. Author has named it **Goby Shark**, because it has a head similar to a shark but the body is more akin to that of a goby fish:

The fossil photos show the **Zamoyski Dragon** appears have a dragon like head and the body of a fish. Three features are visible from the internal view: First, the mouth is actually made up of multiple mandibles. Second, the stomach contents appear to have an unchewed fish being digested in a central vacuole. Third, the upper part of the tail has backward facing barbs.

External View of Zamoyski Dragon - Side A

Internal View of Zamoyski Dragon - Side A

Multiple Mandibles   **Stomach Contents**   **Backward Facing Barbs**

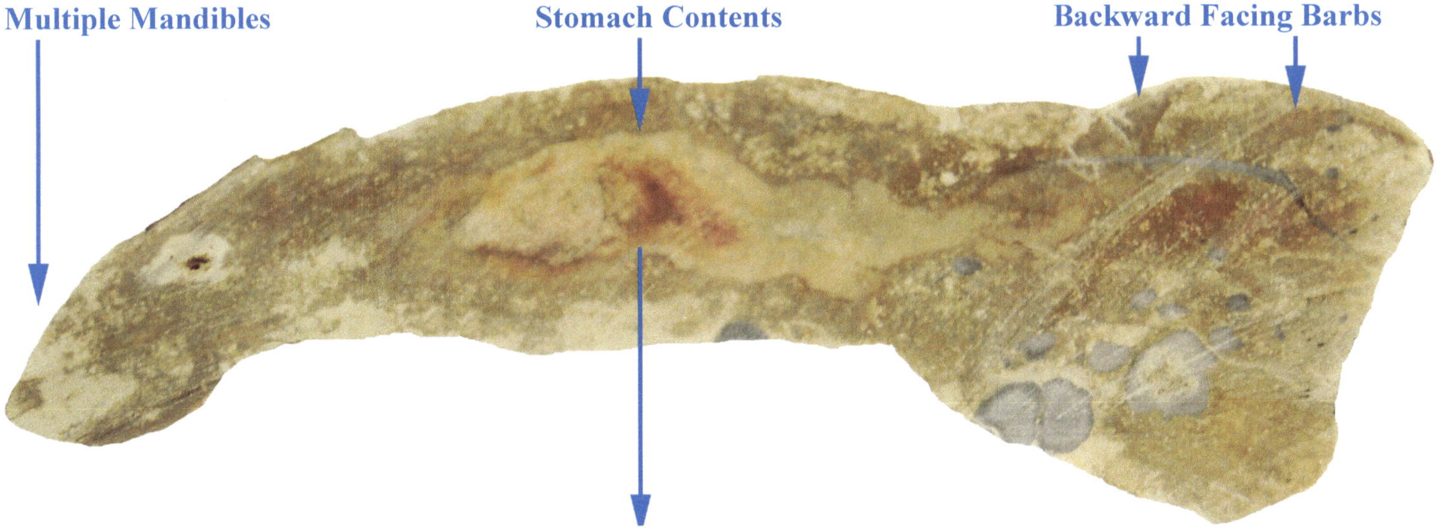

Isolated Stomach Contents (Goby Shark)

The unchewed fish inside (**Goby Shark**) implies the **Zamoyski Dragoon** used a primitive protist strategy of pursue, swallow whole, and digest.

The backward facing barbs on the **Zamoyski Dragon** employ a simple survival advantage in a "pursue and swallow whole" world. The barbs would lodge in the pursuer's throat, preventing swallowing and facilitating forward escape. These barbs would be effective until a "pursue and chew" world arose. The barbs imply the presence of a specialized cell type to make them.

Another example of a hardened exoskeleton is **Mud Worm**, illustrated below:

The actual fossil photo is shown below. **Mud Worm** is ~ 6" (15 cm) long. If its position was accurately preserved in death, it may have lived in mud with only its mouth protruding. The mouth is missing, indicating it was likely made of softer tissue.

Mud Worm
External View

A zoom of the tail shows the visible segmented exoskeleton, including preservation of the color of the likely chitin exoskeleton (i.e. cockroach like color).

Zoom of Tail

Bone

Bone is the most significant development after skin. Bone is a repository for Calcium, Phosphorous, and Mitogens (growth factors). These compounds are routinely moved from extracellular fluid into bone and back from bone into extracellular fluid.

Calcium (Ca++) movement alters nerve function, muscle function, consciousness, and memory. Movement of all three (mitogens, Ca++, phosphorous) enhances activation of the population density management / cell cycle control system. Phosphorous is used in storage of energy (ADP to ATP) and phosphorylation (addition of a phosphorus atom) alters the functions of many proteins.

Movement of these compounds into bone is controlled by a specialized cell called an osteoblast. Osteoblasts also control the population density and activity levels of osteoclasts, a specialized cell type that dissolves bone releasing these compounds back into the extracellular fluid.

Osteoblasts in turn are controlled by numerous endocrines. The result is that many endocrines mediate their effects, in whole or in part, by movement of these compounds into or out of the bone "pantry".

Vitamin D, parathyroid hormone, prostaglandins, and Vitamin A enhance movement from bone into the extracellular fluid. Estrogen, Testosterone, growth hormones (GH, IGF, BMP), and calcitonin enhance movement of these compounds into bone.

As an example, sunlight (UVB) on skin results in synthesis of the **active form of Vitamin D**, which then binds to vitamin D receptors (VDR) in the osteoblasts increasing their production of RANKL (receptor activator of NF-kB ligand) that induces macrophage differentiation into osteoclasts (the bone dissolvers), which in turn results in the release of Ca++, phosphorous, and mitogens from bone into the extracellular fluid. The increased Ca++ in the extracellular fluid **enhances nerve function** by depolarization of nerve membranes (per the Nernst equation), which lowers the threshold required for their firing and enhances neurotransmitter release via the voltage gated Ca++ channels because of both the higher extracellular concentrations of Ca++ and the higher Ca++ concentration gradient differential on the outside of the neuron versus the inside of the neuron. The increased Ca++ **enhances muscle function** by both the enhanced release of neurotransmitter at the neuromuscular junction and by enhanced inrush of Ca++ through the sarcoplasmic reticulum calcium release channels, enhancing Ca++ release into the fluid around the myofibrils, enhancing muscle contractility by removal of the tropomyosin block between actin and myosin, triggering cross-bridge formation and enabling myosin to bind to actin. Increased extracellular Ca++ **enhances brain function** by brain neuron depolarization, enhanced neurotransmitter release, and depolarization of NMDA / glutamate channels to release the Mg++ block, allowing a glutamate mediated influx of Ca++ into the nerve cells and astrocyte mediated amplification of the neuronal transmission that creates the Ca++ wave that underlies consciousness (Periera et. al., 2009) and memory formation (Gibbs et. al. 2009). The release of mitogens and phosphorous into the extracellular fluid **enhances activation of the population density management / cell cycle control system**, with the mitogens binding directly to transmembrane growth factor receptors to initiate the intracellular cascades that transmit the grow and divide signal to alter DNA expression to produce the proteins required for cell division and the phosphorus enhances the many phosphorylations required to transmit the grow and divide signal to the nucleus.

The point being, the integration of bone with soft tissue, is a significant advancement in life forms. Integration of calcium secreting filter feeder DNA, such as that of the Permian Protopharetra shown earlier, may have happened in **Braniac Worm**. The DNA of early CSFF cells may have gone on to become osteoblast cells. An artist rendering of **Braniac Worm** is shown below:

The photos of the actual fossil are shown below. **Braniac Worm** is ~ 5" or ~ 13 cm long and appears to have CSFF vestiges integrated into a flagellate tail.

Braniac Worm
Interior View of Sectioned Side A - Matrix Digitally Excluded

Interior View of Sectioned Side B - Matrix Digitally Excluded

Possible CSFF DNA integrated with flagellate tail

Another possible CSFF / Soft Tissue integration appears in **Bird-Squito** which has a beak like structure. A rendering of **Bird-Squito** is shown below:

The oriented exterior photos of Side A and Side B are shown below. The specimen was cut off-center, resulting in a shorter Side A and longer Side B. The full specimen is ~ 7" long.

A head shot shows a hard skeleton beak, with what appear to be serrations on Side B. The serrations could be for either chewing or be possible filter feeder structures.

External View of Side A

External View of Side B

Serrated Hard Skeleton Beak

Reproduction

Unicellular protozoa have several modes of reproducing: binary fission, asexually, and sexually. An individual protozoan is hermaphroditic.

A possible example of early reproduction appears in **Whaleen**, which was previously reviewed for its optic neuron structure. The illustration and external photo of **Whaleen** are provided first for reference:

Whaleen Illustration

Exterior Photo of Whaleen

It appears **Whaleen's** reproduction may have been hosted in a protective sack in the central food lumen.

Interior Photo of Whaleen

Zoom of Possible Progeny Gestating in Central Food Lumen

Unlike the lumen contents of the **Zamoyski Dragon**, where **Goby Shark** appeared to be in the early stages of digestion, the above possible progeny appears to have been alive and possibly in a protective sack that kept it from being digested.

This protective sack may have been the forerunner of a womb, providing the progeny nutrients directly from the digestive lumen. Eventually, development of a circulatory system allowed the womb to provide nutrients directly from the blood.

Another possible example of reproduction / gestation in the central food lumen, except without the protective sack, appears in a more primitive life form, the **Carnivorous Protopharetra** (~ 6" or 15 cm tall). The life form appears to be an example of soft tissue integrating into a fixed CSFF, resulting in flagellate like flaps at the top of the hard skeleton CSFF structures and possible reproduction by budding in the food lumen.

The artist's illustration of what the life form possibly looked like from the outside is shown below:

Carnivorous Protopharetra

The sectioned fossil photo reveals several internal structures, including the possible budding progeny in the lower part of the food lumen.

Internal View of Sectioned Specimen - Matrix Excluded, Lumen in Gray

Mouth Flaps

Stomach

Possible Budding Progeny

Pre-Triassic / Jurassic Features

Dinosaurs dominated the earth in the Triassic and Jurassic periods that followed the Permian-Triassic extinction event. Dinosaurs first appeared some 230 Ma . The significance of dinosaur type features will be best understood in context of the next chapter .

Geckosaurus (~ 6" or 15 cm tall) has a dinosaur like head (e.g. T-Rex), despite its fairly dainty body.

Actual Headshot of Geckosaurus

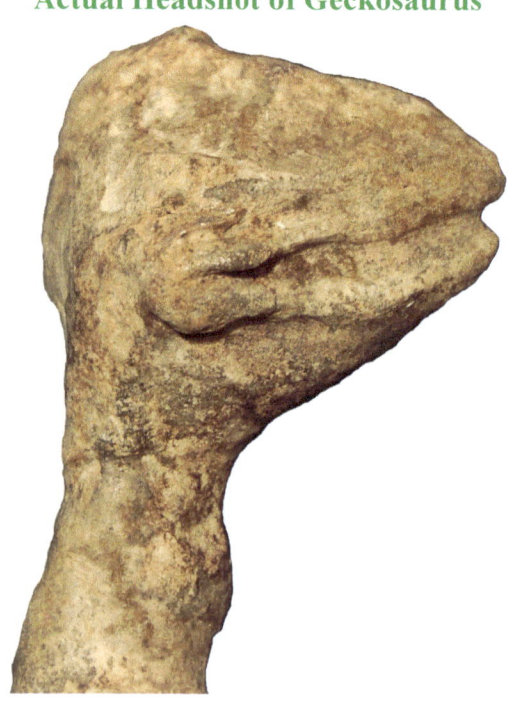

Geckosaurus has a dinosaur like head but a body more akin to a dragon fly or sea horse. A crystal at mid body indicates where a likely flipper or wing forerunner resided.

The interior view shows a jawed mouth and a linear food lumen running through the creature. It is not clear if this is a baby or a mature adult.

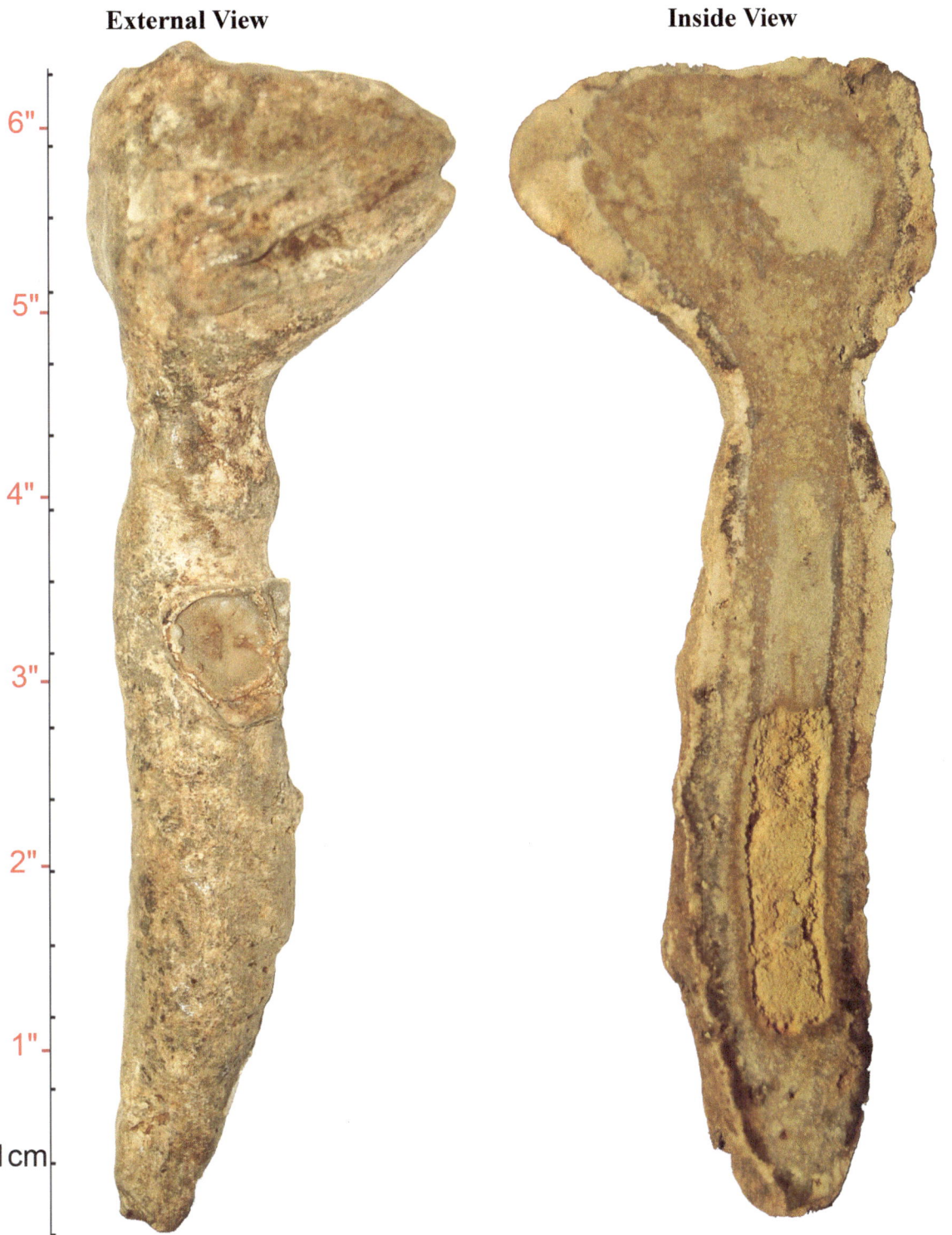

The other specimen with dinosaur like features was the **Zamoyski Dragon**, previously reviewed.

Pre-Terrestrial Features

From a more long term perspective, several traits show up in these marine specimens that have features suited for migration to terrestrial environments.

Claw-Paw Dragon (~ 13" or 33 cm) reveals DNA capable of generating terrestrially suitable pseudopods. An artist rendering of the **Claw-Paw Dragon** is provided below:

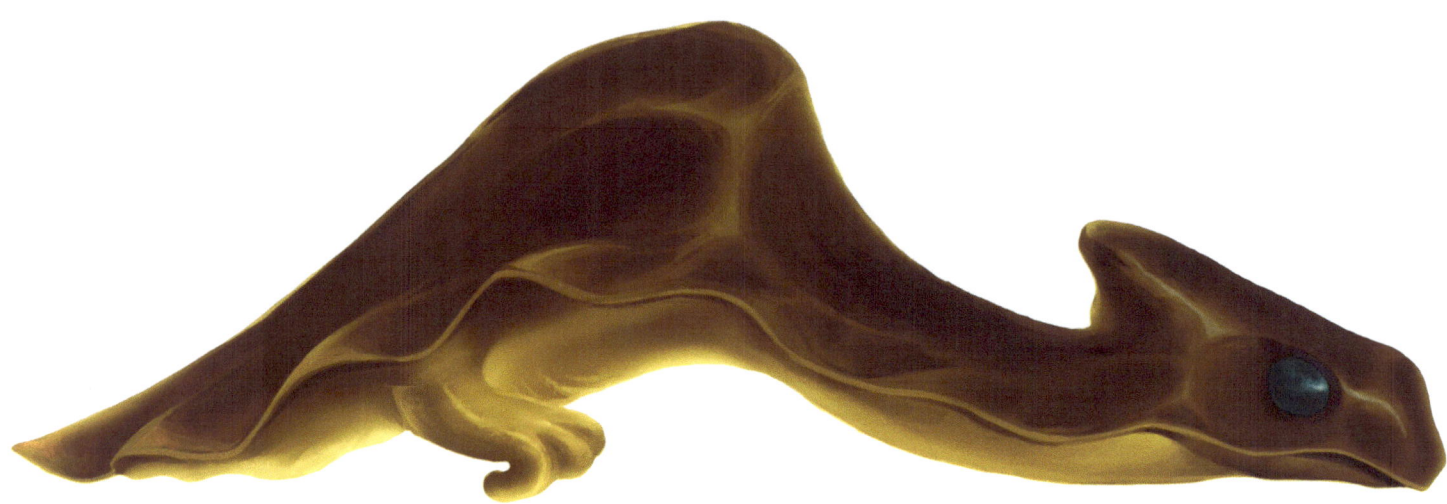

A photo of the actual specimen are presented below:

Claw-Paw Dragon
External View of Sectioned Side A

Rear Paw

Mouth

Zoom of Paw at Rear

Claw-Paw Dragon likely crawled along the sea floor, using its chin as a claw to pull itself and paw at rear to push itself forward. A stump at the top rear of Claw-Paw is likely where a tail resided, but is missing. The tail may have kept Claw-Paw from sinking into the mud as the paw pushed it forward. Claw-Paw dragon also appears to have a breathing orifice that resides above the mouth.

Internal View Side B - Zoom of Breathing Orifice

Breathing Orifice

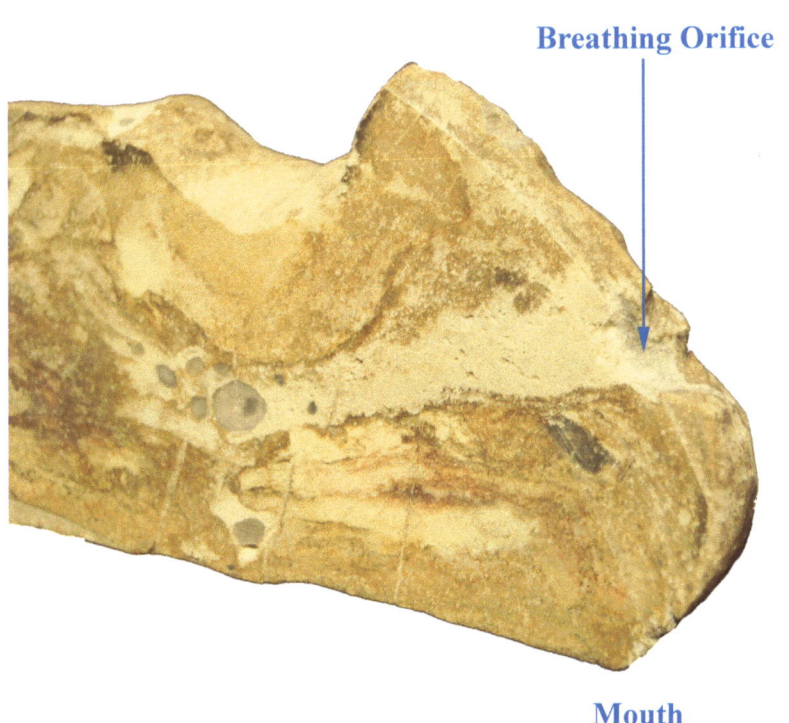

Mouth

A lumen runs along the top and bottom. The upper lumen goes from breathing orifice to the hump. It may have been used to obtain and store oxygen (likely from water, possibly from air).

Claw-Paw Dragon: Internal View of Sectioned Side B

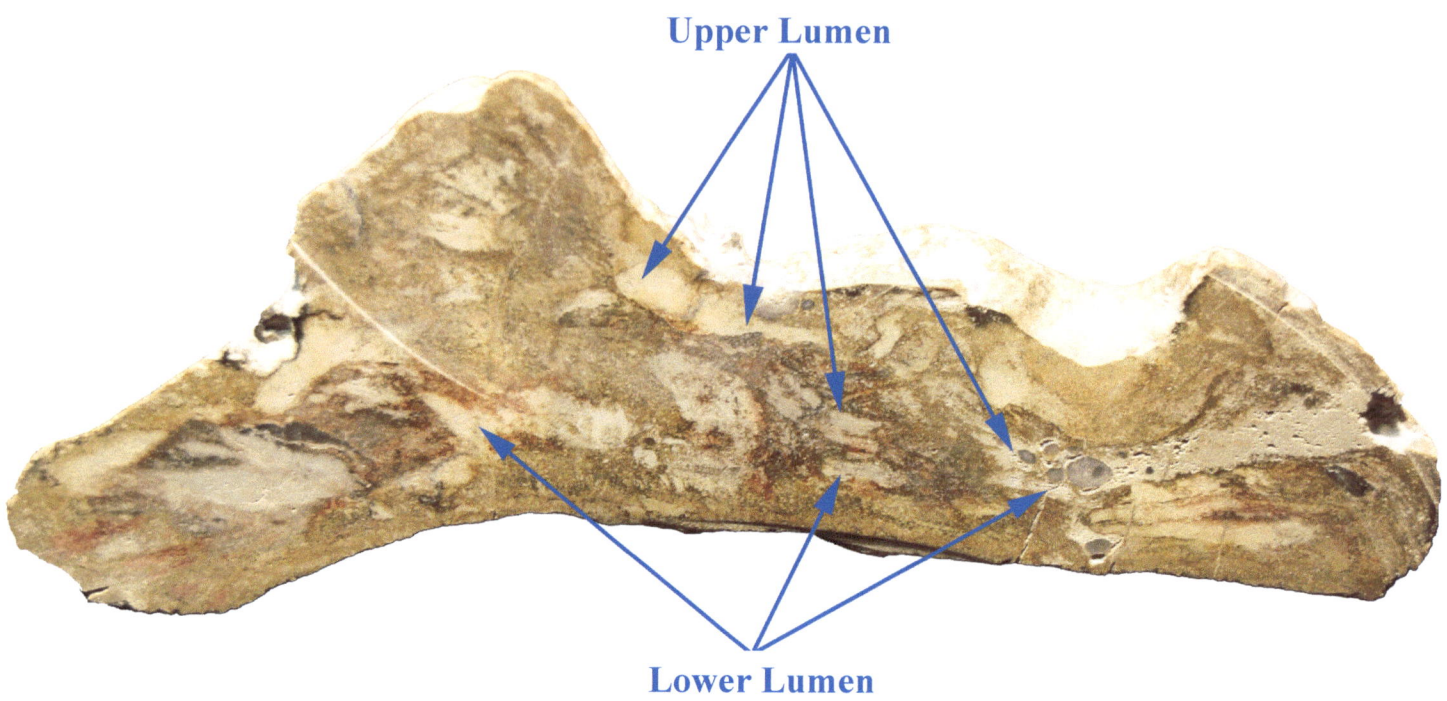

Upper Lumen

Lower Lumen

The lower lumen runs along the bottom and terminates in an anus below the tail. The lower lumen is more typical of a digestive tract.

External View Side A - Zoom of Anus

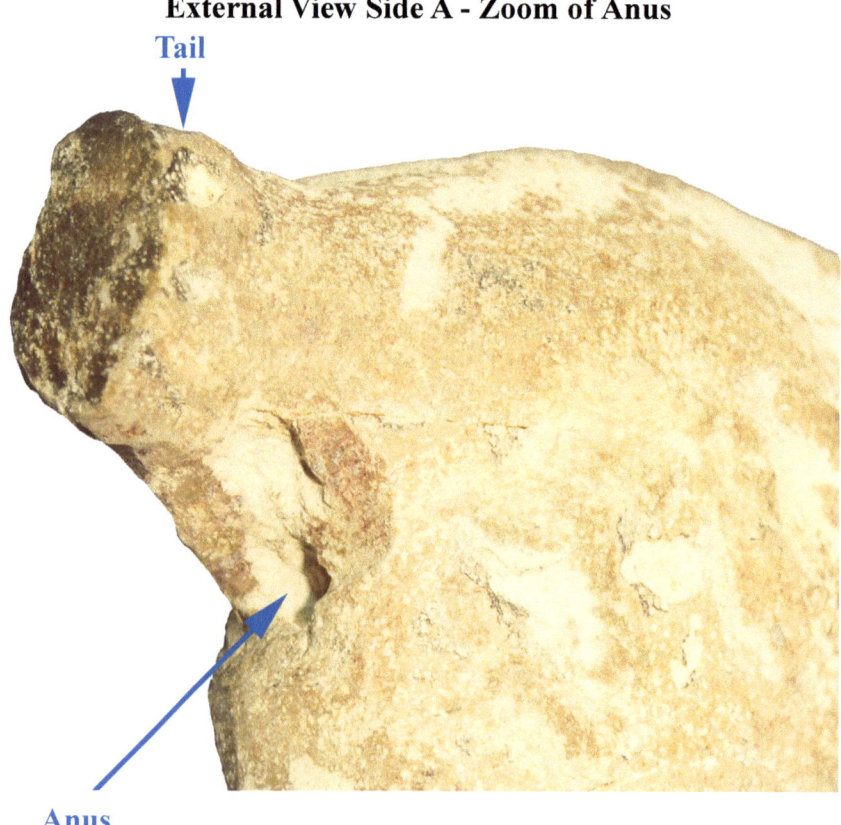

Tail

Anus

A more significant example of a pre-terrestrial life form is the Arizona Lung Fish (~ 5.5" long). The artist's rendering of the Arizona Lung Fish is shown below:

The fossil photo, compared to a current day lung fish, is shown below:

Arizona Lung Fish
Interior View of Sectioned Specimen Compared to Current Day Lung Fish

Lung fish are believed to be forerunners of terrestrial tetrapods. Their lung is a modified swim bladder that can absorb oxygen and remove waste. Its 4 limbs are in the same relative position as those of terrestrial tetrapods.

Lung fish use aestivation (dormancy) burrows when their pools dry up. They breathe air through their swim bladders and drop their metabolic rate. Lung fish aestivation burrows are found in the Devonian, Carboniferous, and Permian periods (~ 400 Ma - 250 Ma), tending to indicate their significance and survival throughout history. It also implies marine DNA is readily capable of spawning terrestrial life.

Three genera of lungfish are still alive today, one on each of three continents (Australia, South America, and Africa).

Autopsy Summary

The DNA reassortment (aka origin of life, aka biodiversity explosion) event appears to have resulted in altered unicellular DNA that created unicellular giants and also created simple life forms comprised of two or more cell types. These early multicellular life forms have morphological features similar to those seen in life forms today. Their absence of proper bones or vertebrae, however, indicates they may be an intermediate step between unicellular life and vertebrate multicellular life.

Based on the autopsy results and molecular biology of DNA expression previously discussed, we can start looking for perpetrators that had the means, motive, and opportunity to commit this DNA reassortment. We will start with the suspect identified at the very beginning of this chapter, the **Permian Protopharetra**, that was also seen at the scene of the Cambrian biodiversity explosion event.

That's where we are off to in the next chapter.

Chapter 3

The Perpetrators of Multicellular Life

We now want to see who had the means, motive, and opportunity to commit this DNA reassortment.

Opportunity

The **Permian Protopharetra** presented in the last chapter was also spotted at the scene of the Cambrian "explosion of life" event, so let's take a closer look at this suspect.

As previously mentioned, our suspect is an ancient single cell eukaryote that lived in colonies and is the first known organism to have secreted calcium carbonate as a skeletal material. They were part of a crime family we will call Calcium Secreting Filter Feeders (CSFFs), that channeled moving ocean water through a hard skeleton labyrinth, to obtain nutrients suspended in water.

CSFFs left reeflike deposits during both the Cambrian explosion (*Boardman, Fossil Invertebrates, 1987, p.107*) and post Permian explosion (*Science Daily, Oct. 5, 2011*). That prominently places the CSFFs at the scene of both explosion of life events.

That establishes they had the opportunity.

Motive

Their motive for channeling moving water through a hard skeleton labyrinth was to obtain food.

Means

To see if they had the means to commit random DNA reassortment of cells suspended in this moving water we need to combine fluid dynamics with molecular biology. This whole creation of multicellular life thing may have simply been an unintended consequence of the way they ate.

In Chapter 1 we discussed some differences between prokaryotic and eukaryotic cells, such as prokaryotic cells having circular DNA and eukaryotic cells having linear DNA (much of which was not used), eukaryotic mitochondria being of prokaryotic origin, and the differences in cell wall construction of the two cell types.

Additional information on cell walls and membranes is necessary in order to properly understand the evidence about to be presented.

Prokaryotic cells have a rigid, triple layer cell wall structure. Eukaryotic cells have a flexible, single lipid bilayer membrane that is a subset of a prokaryotic cell wall, shown diagrammatically below:

Cross Section of Cell Wall

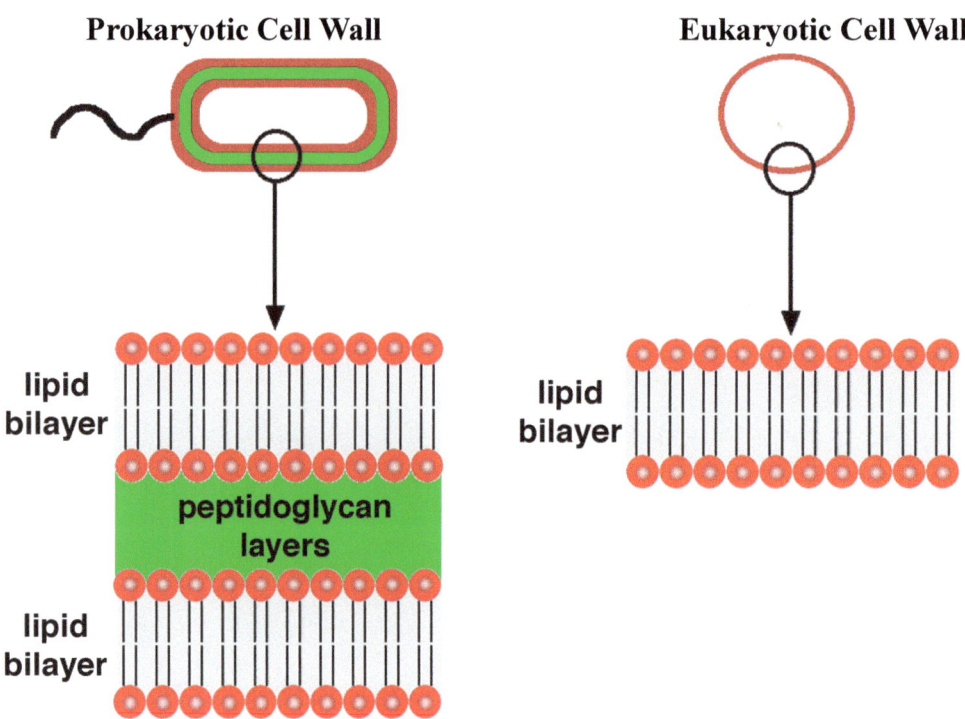

Lipid bilayers are made up of molecules that have a water loving head (hydrophilic) and lipid loving tail (lipophilic). When placed in water, they self assemble to form compartments. Likewise, if a lipid bilayer of the prokaryotic wall shown above was scraped off in water, it would self assemble into a lipid bilayer compartment.

Self Assemble in Water

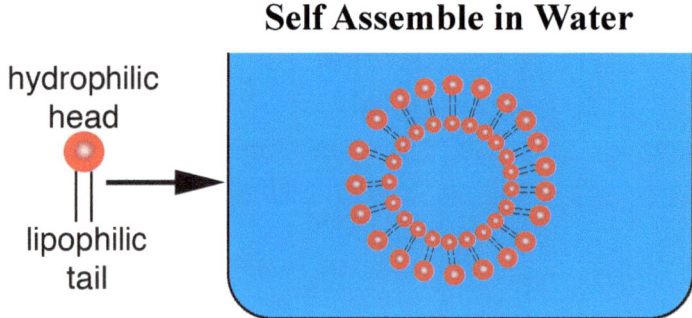

Internal Membrane Compartments

Prokaryotes lack internal membrane bound compartments. Eukaryotes use internal membrane bound compartments (nucleus, mitochondria, Golgi apparatus). The membranes are made of lipid bilayers.

Mechanistic Principles for Creating "Grab Bag" DNA Reassortment

If one desired to create new eukaryotic cells with enormous potential biodiversity, it would require only three conditions.

1) Cells aggregated in close proximity to each other in water:

2) Shear forces or structures capable of rupturing cell membranes:

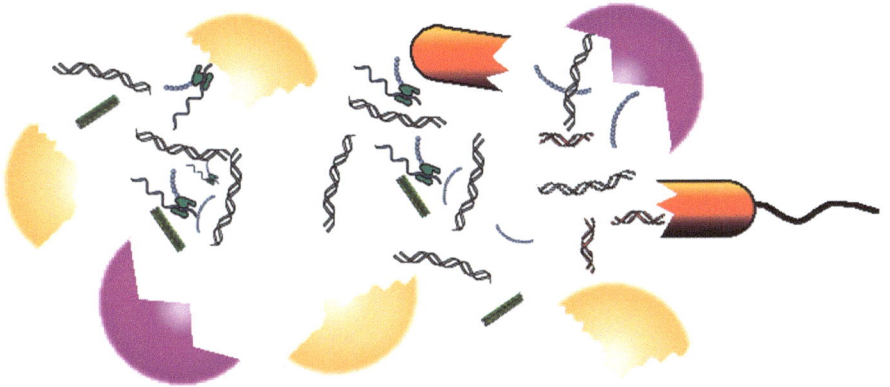

3) A confined space where the spontaneously reassembling lipid bilayers could effectively encapsulate a batch of the ambient genetic slurry.

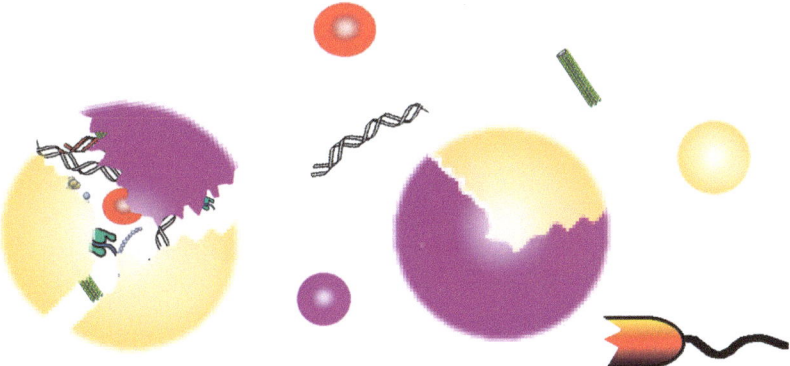

The "grab bag" or random DNA reassortment process could be expected to generate cells with much lower genomic efficiency than the cells one started out with, as well as having much unused DNA.

Although both prokaryotic and eukaryotic cells could be used as input into the process, only eukaryotic cells would emerge as output of the process, because of the ability of their cell membranes to self assemble.

The resulting eukaryotic cells could also be expected to have membrane bound compartments inside the main cell wall membrane compartment.

Review of 4 CSFFs - Did they have the means?

Coastal oceans have around 1,000,000 suspended cells per ml of water. To obtain the intracellular nutrients, such as proteins and nucleotides, the cell wall would need to be sheared open. This is the presumed motive for channeling water through the hard skeleton labyrinth structure.

But are these structures also capable of creating cells with reassorted DNA?

We now examine 4 bodies of Calcium Secreting Filter Feeders (CSFFs) found at the scene to see if they have structures capable of achieving the "grab bag" DNA reassortment process previously described.

Body 1 is the "Permian Protopharetra" previously shown.

Permian Protopharetra

Internal View, in Matrix - Side A   **Internal View, in Matrix - Side B**

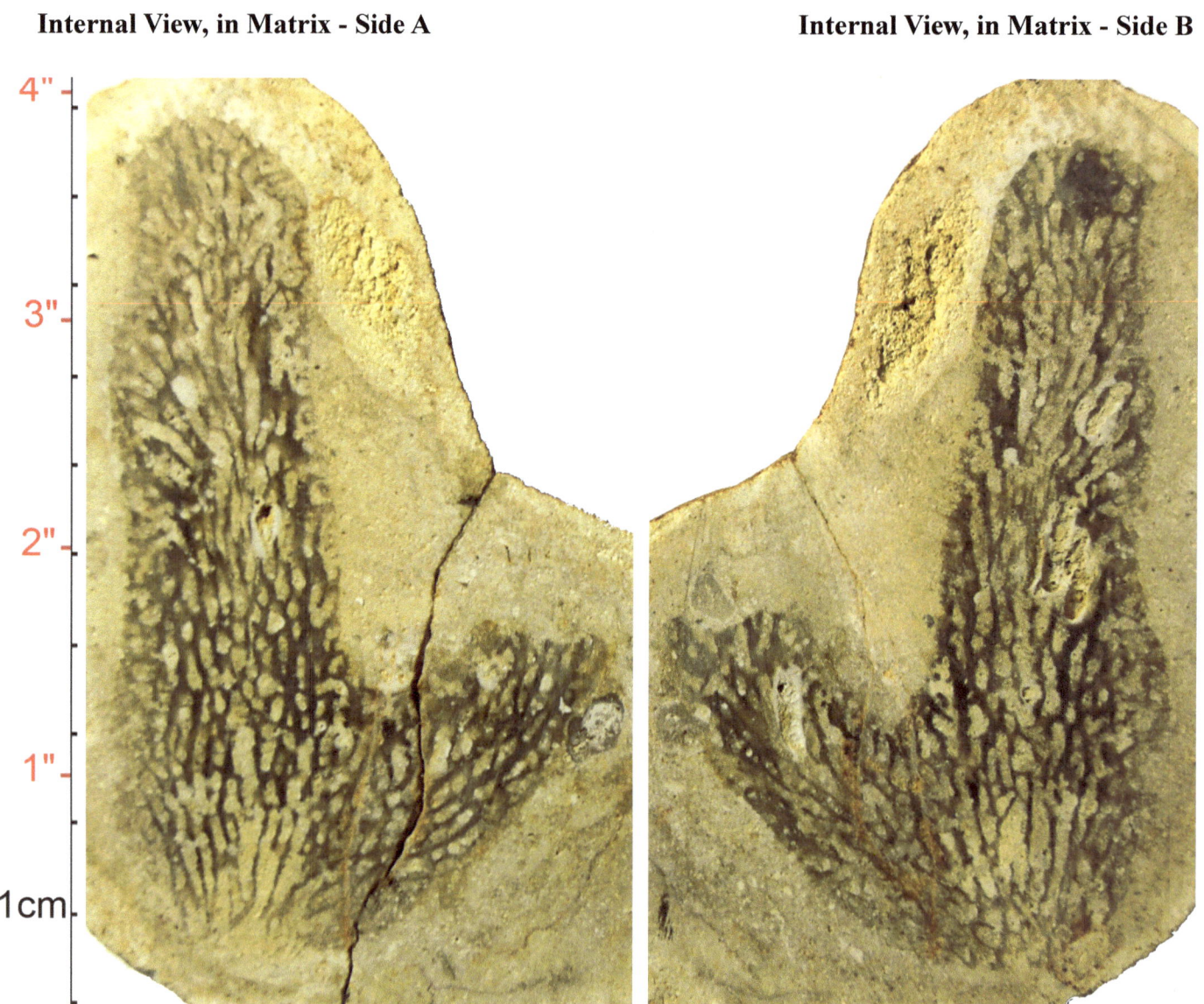

Body 2 is of a **Spherical CSFF**. The artist's rendering and actual photos are shown below:

Spherical CSFF

Internal View, in Matrix - Side A   **Internal View, in Matrix - Side B**

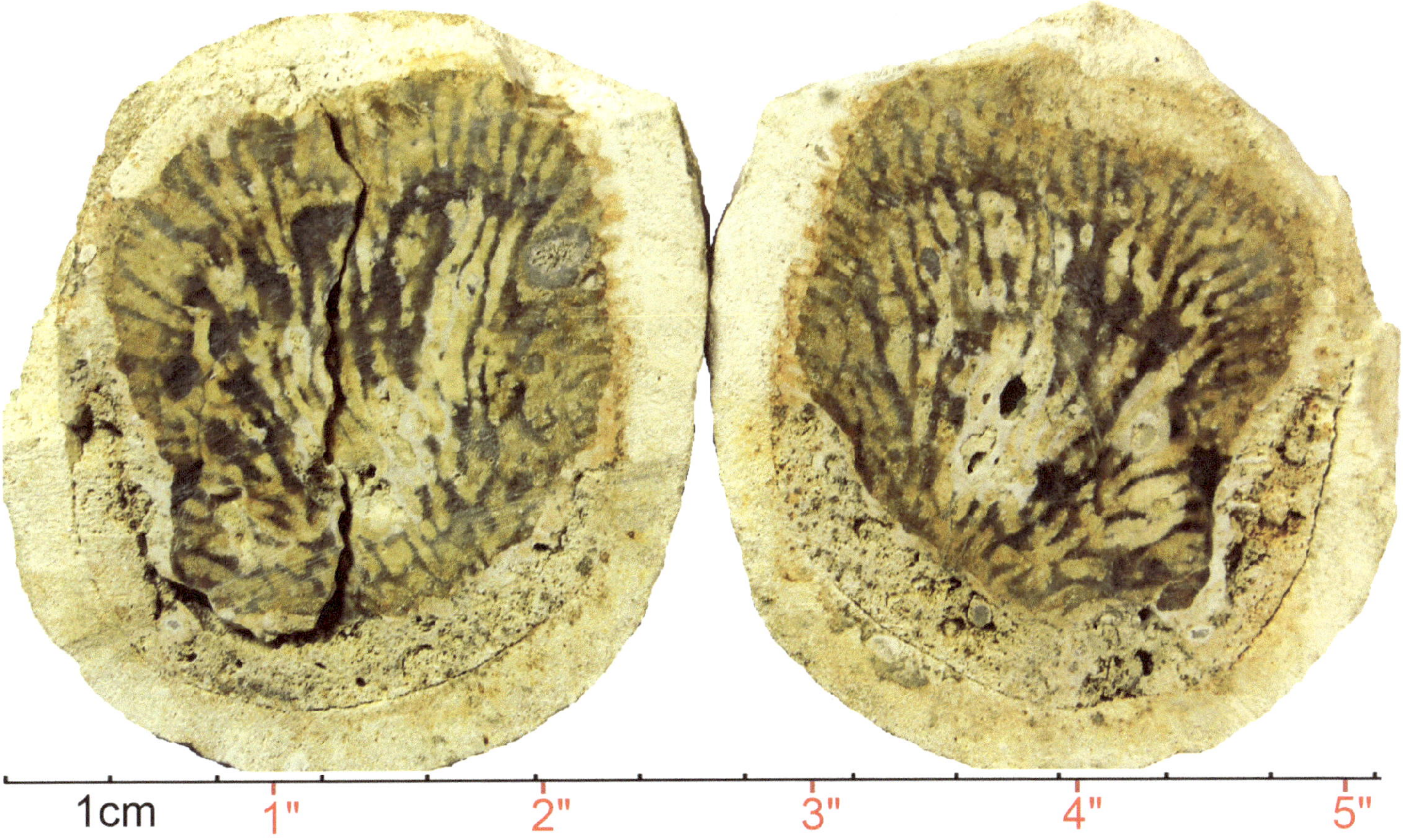

1cm   1"   2"   3"   4"   5"

Body 3 is called **Mega Protopharetra**.

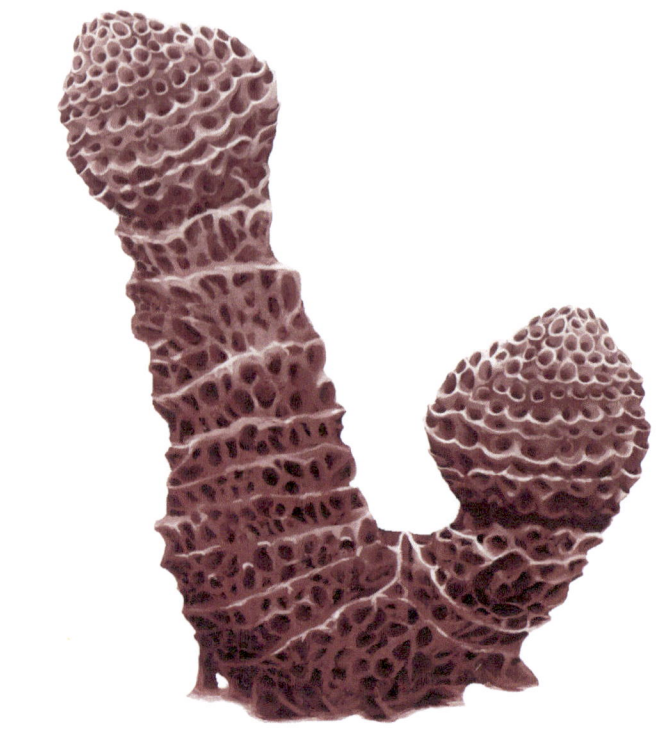

Internal View, in Matrix - Side A

Body 4 is of a **Columnar CSFF** shown in the Illustration and Photos below:

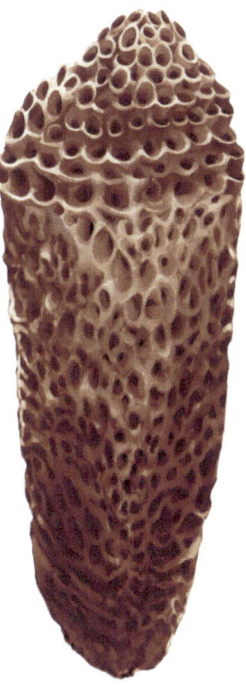

Internal View, in Matrix- Side A **Internal View, in Matrix - Side B**

Fluid Dynamics Meets Molecular Biology

A review of the fluid dynamics of several internal CSFF structures, combined with molecular biology, is covered next, starting with one of the structures visible in the **Spherical CSFF**.

Structure 1: Nozzle and Slurry Chamber Structure

Enlarged View of Structure   **Lumen / Water Channel in Blue**

The dark parts on the left outline the hard skeleton structure and the lumen is filled with the lighter colored calcium matrix . The photo (right) has the lumen / water channel in blue for clarity.

A tracing of the structure is shown below. Two noteworthy attributes are: 1) a nozzle structure that amplifies water velocity and 2) a post nozzle structure (slurry chamber) that enhances turbulence.

Diagrammatic Depiction of Structure's Fluid Dynamics

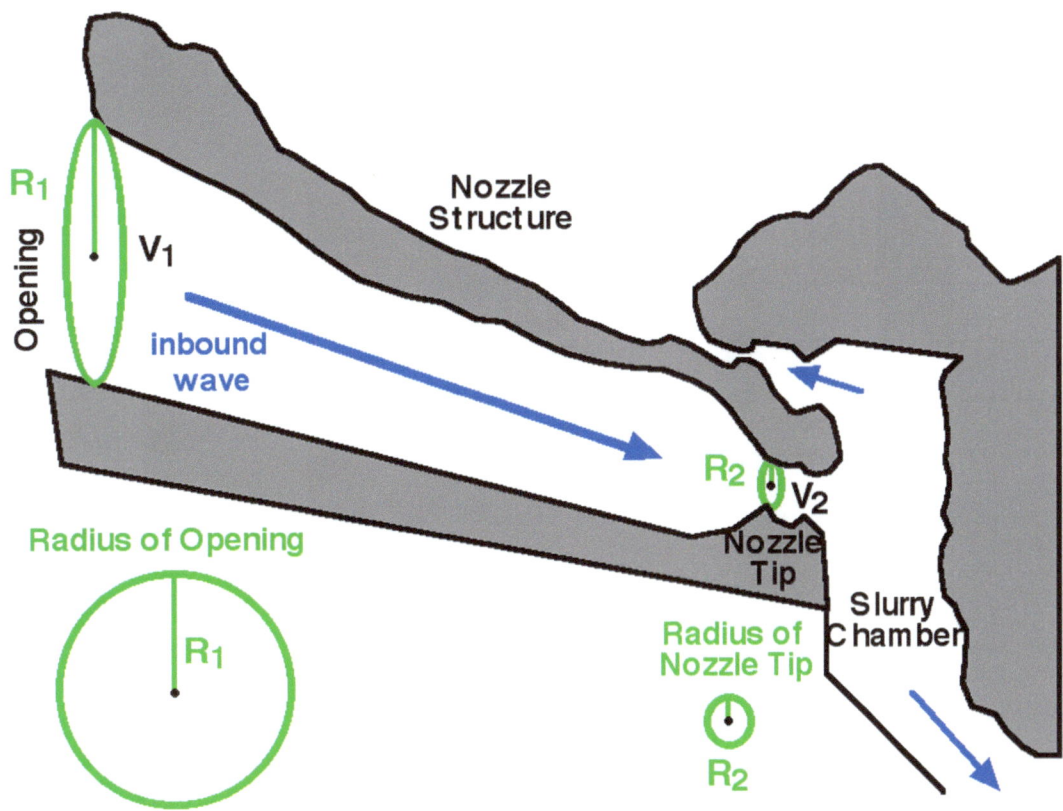

Water Velocity Amplification (~16X): The opening on the ocean side of the nozzle orifice is more than 4 times larger than the nozzle tip that feeds the slurry chamber. For a given flow (e.g. in cubic mm / sec) coming in from the ocean, water velocity increases exponentially as the water passes through a confined space. The reason is that flow (**Q**) equals the velocity (**V**) times the cross sectional area (**A**) or **Q= VA**. The area (A) of a circle is **A= πR^2** where R is the radius and π = 3.14. For a given Q, a reduction in radius results in an exponential reduction in area (i.e. R^2), which in turn requires an exponential increase in velocity to maintain equality.

For a circular pipe: $\mathbf{Q = \pi R^2 V}$ or $\mathbf{V = Q / \pi R^2}$

For a given Q, velocity at the opening is $\mathbf{V_1 = Q / \pi R_1{}^2}$ and velocity at the nozzle tip is $\mathbf{V_2 = Q / \pi R_2{}^2}$

Accordingly, for a given Q, the velocity will increase exponentially with a reduction in radius. If the radius is reduced 4 fold, the velocity will increase 16 fold (i.e. 4^2).

The 16 fold velocity amplification shooting out of the nozzle tip into the perpendicular wall is analogous to accelerating a car from 10 mph to 160 mph by the time it hits the brick wall.

Turbulence: Turbulence occurs when a high velocity stream of water enters low or no velocity water. Vortexes form at the border region of the two bodies of water. The vortexes spin water backwards and perpendicular relative to the direction of the high velocity stream. This can be thought of as "nature's blender".

An example of turbulence, in an occluded blood vessel, is shown below. As blood, with its suspended cells, is squeezed through the occlusion, it undergoes velocity acceleration (from the equations above). As it enters the lower velocity blood past the occlusion, turbulence results. Even though blood vessels are soft and blood velocity is low, damage to cells from this process results in a higher risk of stroke.

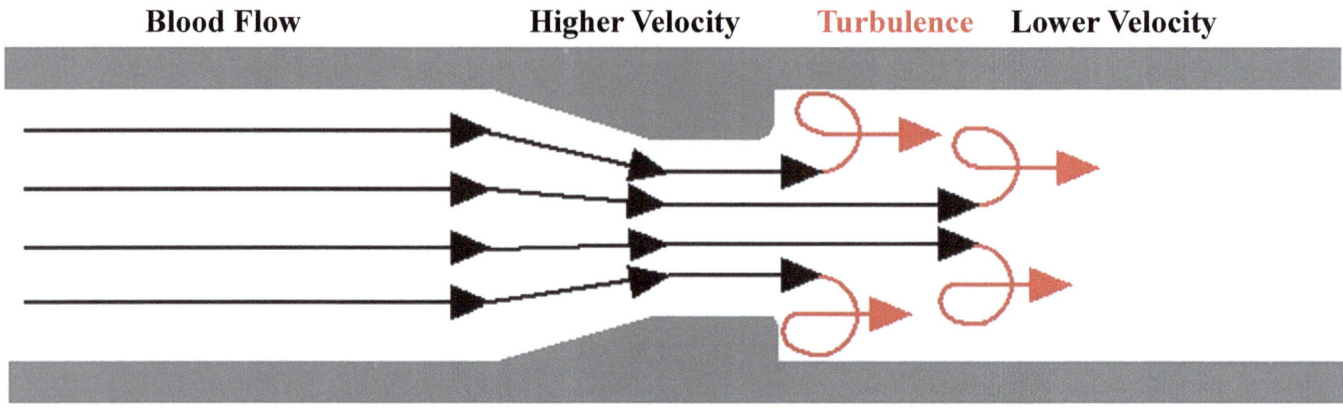

Occlusion

Boosting velocity and replacing the soft blood vessel with a hard skeleton labyrinth, we can begin to understand what happens in a CSFF. In the CSFF under consideration, an **inbound wave (blue)** is accelerated to high velocity through the nozzle tip and hits the back wall of the slurry chamber. **Vortexes or turbulence (red)** could be expected to form in several places as this high velocity stream enters and travels through the no or low velocity water in the slurry chamber. Some of these anticipated turbulence zones are shown below in red:

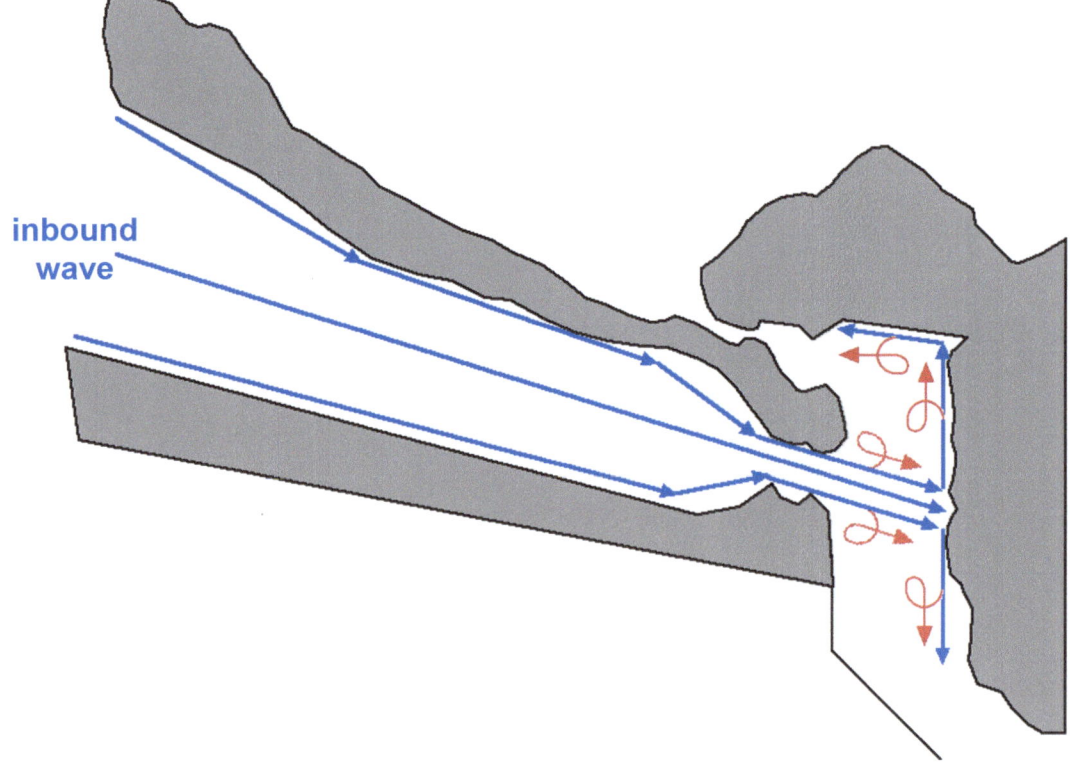

Combining fluid dynamics with molecular biology we can now review the structure's mechanism of action to see if it has the means to achieve the proposed genetic reassortment.

Diagrammatic Depiction of Nozzle and Slurry Chamber Structure in Action

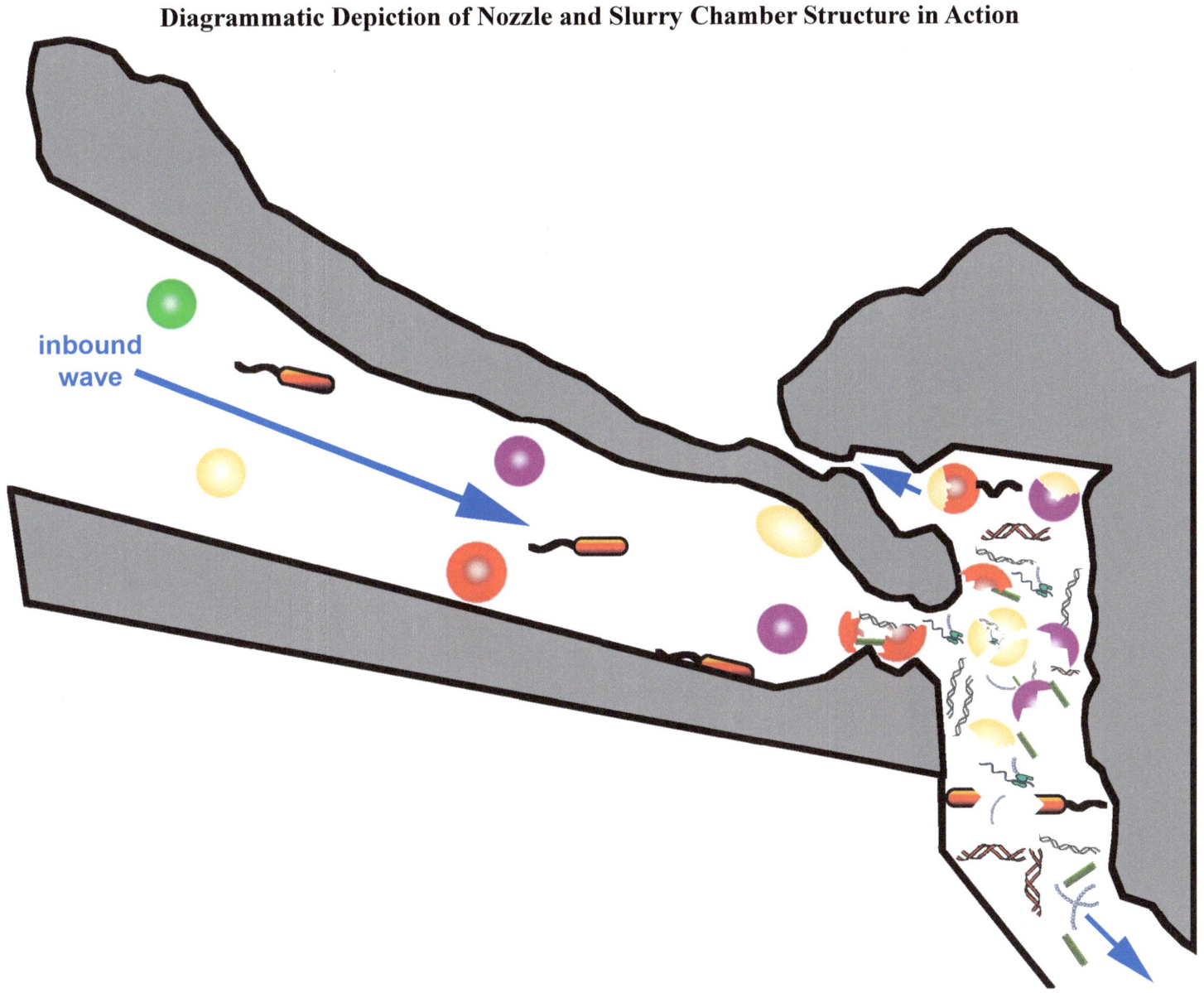

An inbound wave with its 1,000,000 suspended cells per ml is accelerated 16 fold and smashed into a perpendicular hard skeleton wall (back of the slurry chamber).

The ruptured cells in the slurry chamber are subjected to turbulence as the high velocity stream enters and travels through the low (or no) velocity water in the slurry chamber.

As the lighter lipid membranes spontaneously assemble, they take a gulp of the genetic slurry and can escape through the top vent. Heavier proteins and DNA fragments settle downwards, presumably toward the feeding colony.

The structure fulfills the requirements for grab bag reassortment. This suspect had the means.

Structure 2: Cyclone Structure

A close up of the top right branch of the **Mega Protopharetra** shows two conjoined structures that resemble two simple cyclonic filters. The extended common center wall would enable cyclonic filtering of both inbound and outbound waves.

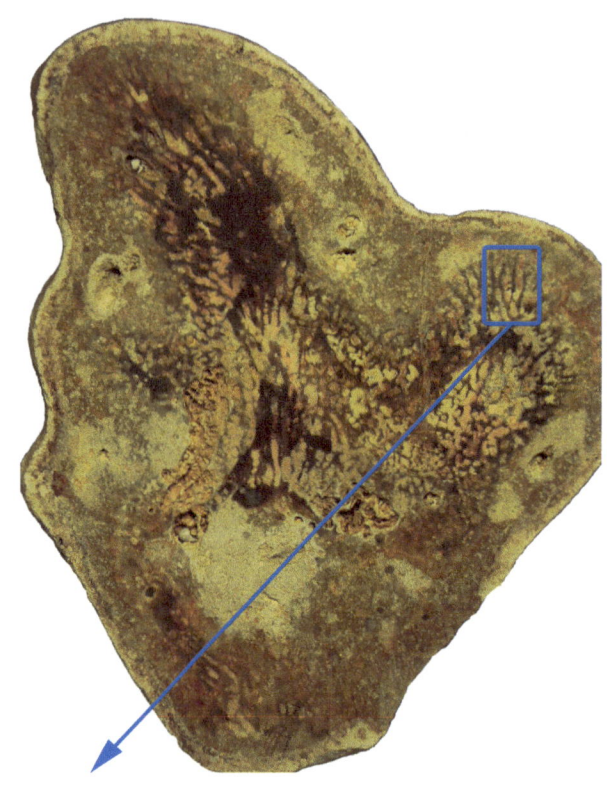

Zoom	Cavities in Blue

Diagrammatic Depiction of Cyclone in Action

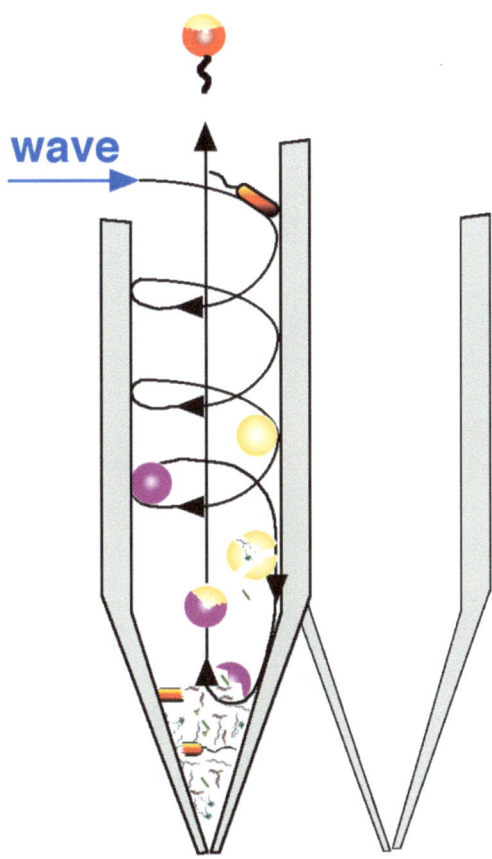

A simple cyclone operates when a perpendicular stream of air (or water in this case) is introduced against the walls of the cylinder. The result is that the stream spirals down the wall, centrifuging larger suspended particles against the walls. As the cyclonic structures taper, the area is reduced and water velocity exponentially increases (from the velocity and area equations perviously discussed). As the stream reaches the bottom part of the filter, it begins to flow upward through the center of the spiral and exits the cyclonic structure.

The heavier centrifuged particles collect at the bottom of the cone. This is basically how a household cyclonic vacuum cleaner works.

As an inbound wave enters, cells are centrifuged against the walls as the stream spirals down the cyclone. This is analogous to pressing down and dragging a grape over coarse sandpaper. The cell membranes would be sheared and the intracellular nutrients released.

While the heavier, less buoyant DNA, RNA, and Proteins collect at the bottom of the cone, the lighter, more buoyant disintegrated lipid membranes reaching the bottom reversal point of the stream would have the advantage of being propelled through the slurry, taking a gulp of the genetic slurry as they reassemble. A portion of the lighter reassembled cells would hitch a ride on the upward stream and be ejected out of the cyclone filter.

Cyclone structures appear in all four CSFF specimens. They appear to be the preferred structure as various modifications are employed to allow extensive use of cyclonic activity. The above dual cyclone structure appears to have the advantage of cyclonic filtering of both inbound and outbound (ebbing) waves.

Structure 3: Nozzle / Cyclone Structure

Both the **Permian Protopharetra** (shown below) and **Columnar CSFF** specimens contain a Nozzle / Cyclone Combination Structure. The cyclone is prefaced with a nozzle to accelerate water into the cyclone (presumably for enhanced shearing of cell walls), with the accelerated water injected perpendicularly into the cyclone wall. A collection / slurry chamber is located at the bottom of the cyclone.

Permian Protopharetra

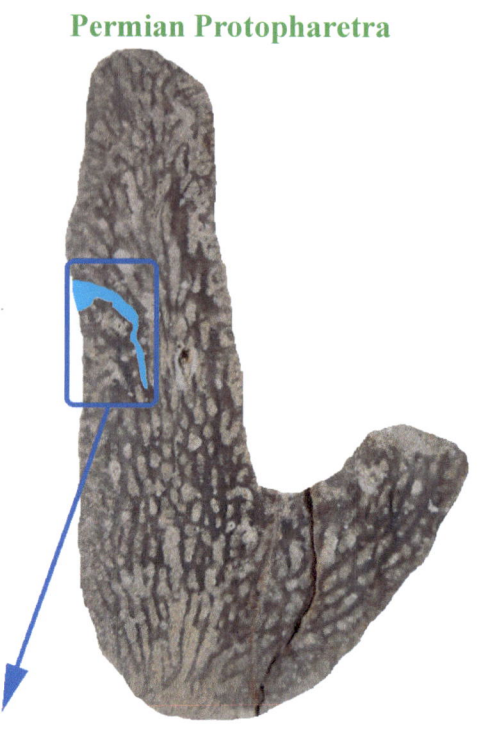

Enlarged Photo, Cavity in Blue

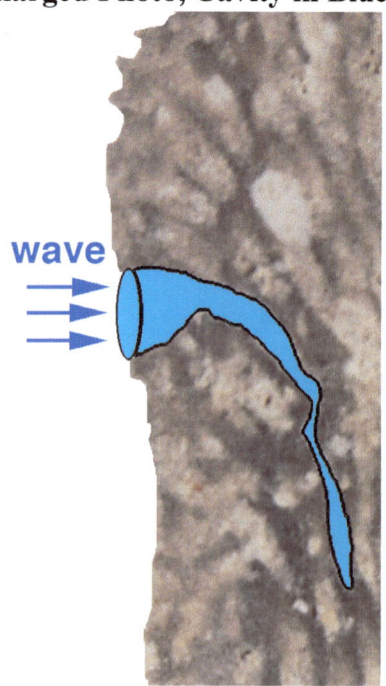

Diagrammatic Depiction of Structure in Action

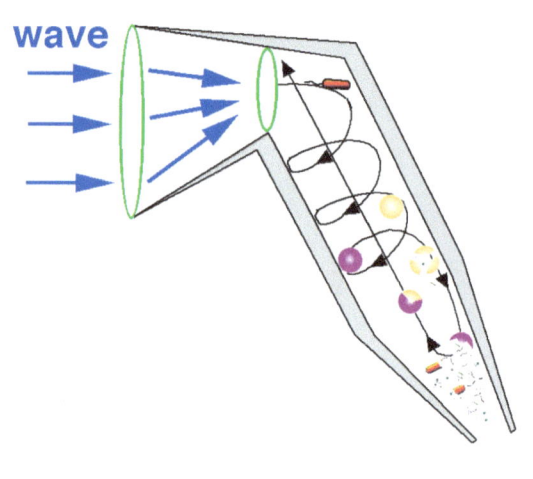

Structure 4: Nozzle / Cyclone / Nozzle Combination with Slurry Chamber

The Columnar CSFF shows a Nozzle / Cyclone / Nozzle combination with a Slurry Chamber, shown in blue in the photo of **Structure 4** below. The lower inlet **(B)** has a larger diameter opening which tapers (i.e. a nozzle structure) and injects water into a cyclone structure. The cyclone in turn tapers at its end into a nozzle tip before injecting the stream into the slurry chamber.

Structure 5: Dual Port Injection Structure

A fifth type of structure observed (in the Permian Protopharetra) is a dual port water entry structure (**Structure 5** below). It merges two streams **(A and B)** into a single structure, with one of the streams being perpendicular to the other. If the lower stream **(B)** is offset from center point of the upper stream **(A),** so as to hit the upper stream at its periphery, it could potentially be used to spin the upper stream.

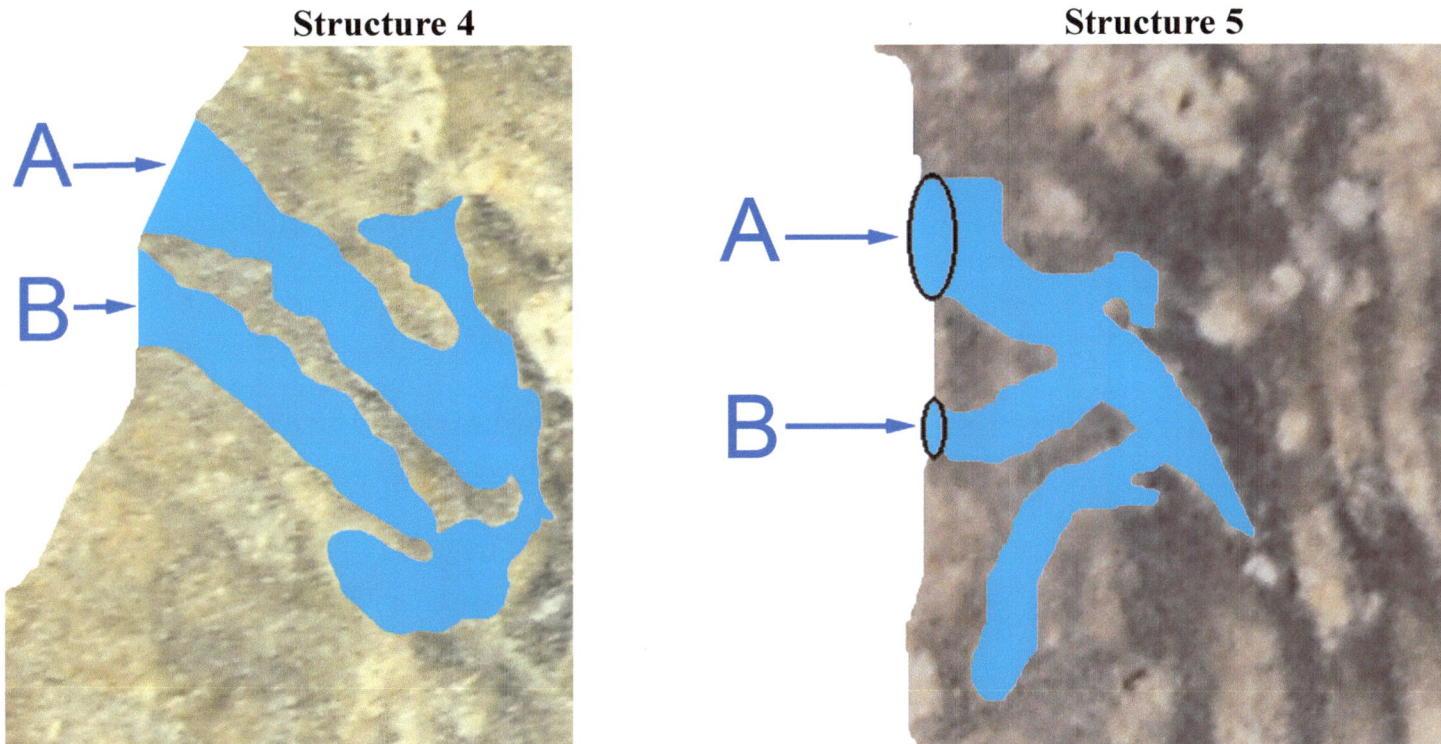

| Structure 4 | Structure 5 |

In summary, the common goal of all of these structures appears to be to create accelerated, turbulent streams inside a hard skeleton labyrinth. The purpose apparently was to shear membranes of suspended cells to obtain intracellular nutrients. The unintended consequence was a genetic reassortment process capable of creating unimaginably biodiverse life forms.

The above examples presented are likely only a subset of such structures that have existed through time. Many CSFF structures can be found throughout the fossil record and many likely contained structures capable of hosting genetic reassortment.

In contrast, CSFFs alive today, such as corals, have an organism that lives in a small cup or channel at the outermost part of the exoskeleton. The organism actively snatches food as it passes by, and no water is accelerated through a hard skeleton labyrinth.

Consistency of the Process with Observed Molecular Biology

The proposed genetic reassortment process would account for transfer of prokaryotic characteristics to eukaryotic cells and the explosion of multicellular eukaryotic life. It is consistent with observed molecular biology in the several areas reviewed.

First, the process would preferentially favor eukaryotic cell development, as bacterial cell walls cannot spontaneously reassemble. This is consistent with the explosion of eukaryotic life.

Second, this process would also account for how the transfer of the mitochondrial DNA from prokaryotic cells to eukaryotic cells occurred.

Third, the lipid bilayers used for internal compartments by eukaryotic cells are also consistent with the process proposed above. Smaller reassembled lipid bilayer encapsulated structures, such as mitochondria and the nucleus, could have also been encapsulated within a larger reassembling lipid bilayer structure.

Fourth, the process would account for much lower genomic efficiency of eukaryotic cells. The eukaryotic genetic junkyard is consistent with a "grab bag" reassortment process.

Fifth, the process would account for the explosion of biodiversity. The small percentage of viable cells created would require this process be conducted on an extremely large scale. Reef like deposits left during both the Cambrian and Permian/Triassic biodiversity explosion events are a testament to the scale on which CSFFs operated to generate enough viable life to establish a new, stable ecosystem.

As more complex, reassorted eukaryotic cells became available in the environment, the subsequent genetic reassortment product could yield progressively more complex life forms.

Eventually, establishment of predators would tend to limit further biodiversity and establish a stable ecosystem. As exciting as a new viable life form could be, it would only be lunch for a predator.

Chapter Summary:

The CSFFs had the opportunity and means to perpetrate the DNA reassortment. The output of the reassortment is consistent with both known molecular biology and the autopsy results from the prior chapter. The motive was to obtain intracellular nutrients. Reassortment was an unintended consequence.

The disappearance of the CSFFs as a new ecosystem emerges has also been a mystery. Some of the bodies appear to provide insight into the demise of the CSFFs and are covered in the next chapter.

Chapter 4

The Fate of the Perpetrators

As abruptly as the CSFFs appear, they once again disappear as a new ecosystem emerges. This appears to have happened in both the Cambrian and Permian explosions of life.

Six CSFF bodies from the Permian site offer some clues as to the possible cause for their demise.

First, the fixed nature of the CSFF (lack of motility) dooms them when events such as tectonic activity lifts the ocean floor, leaving the reef high and dry.

Second, parasitic relations appear common, leading to the demise of the CSFFs. With the large amount of collected intracellular nutrients, CSFFs are akin to a honey comb full of honey, and an easy target for parasitic invasions. CSFFs occasionally also appear to function as incubators for the cells they spawned.

Third, even relations that are symbiotic, appear to alter the nature of the CSFFs so they no longer perform the grab bag DNA reassortment function. Starting with the symbiotic relations, the first example is the **Carnivorous Protopharetra**, previously shown in Chapter 2.

Illustration	Photo of Sectioned Specimen

The rounded bottom indicates it was still attached to a rock like a typical CSFF.

The top part appears to have integrated with soft tissue. Soft tissue covering the inlets would prevent channeling of ocean water though a hard skeleton labyrinth. The resulting organism would likely no longer be able to perform the grab bag reassortment process.

As another example, a Columnar CSFF appears to have integrated with soft tissue to become a **Carnivorous Columnar CSFF**.

In contrast to the **Carnivorous Protopharetra** above, which appears to possibly exhibit colony like behavior, the **Carnivorous Columnar CSFF** appears to be a single organism with the soft tissue adapted for active capture of food.

Columnar CSFF   **Carnivorous Columnar CSFF**

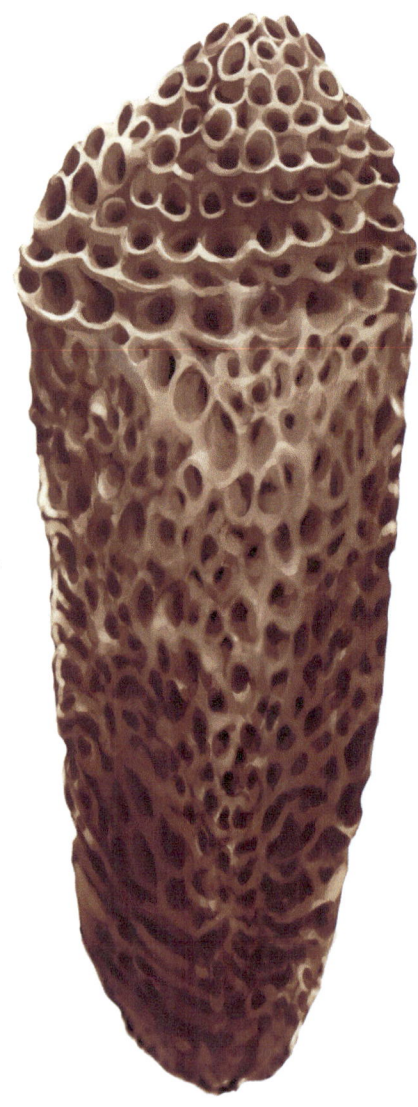 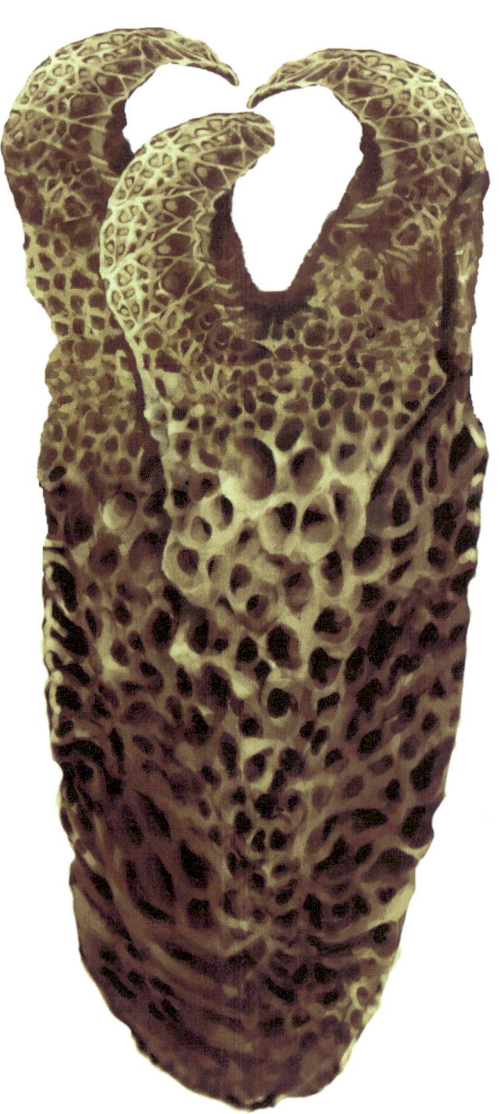

The actual photo of the sectioned **Carnivorous Columnar CSFF** is shown below:

Carnivorous Columnar CSFF

**Sectioned Side A
In Matrix**   **Sectioned Side A
Matrix Digitally Excluded**

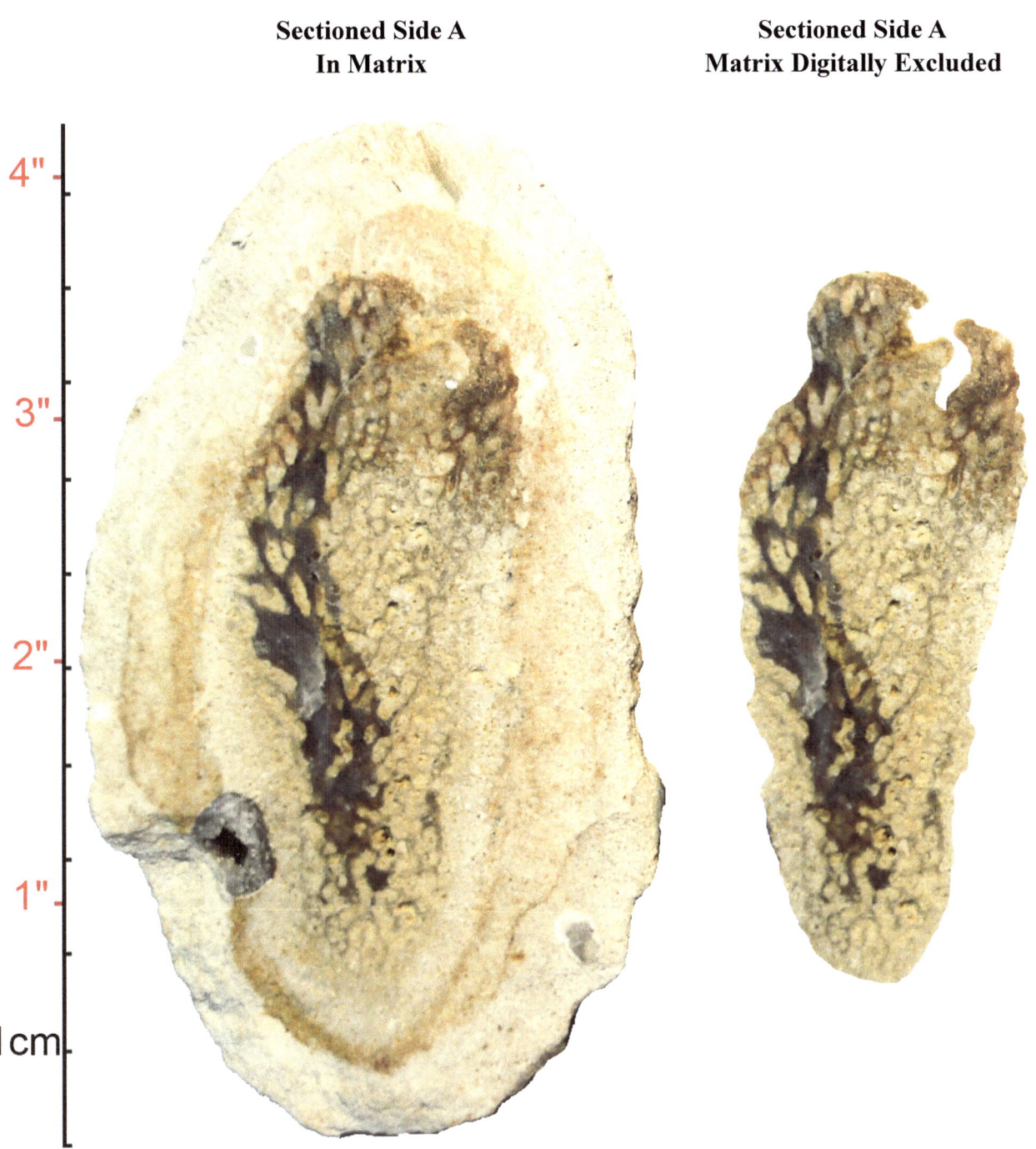

As these CSFFs progressed to active capture of larger prey, the resulting organisms above would no longer perform the grab bag reassortment process.

An example of a parasitic relation appears to be the **Large Encased CSFF**. An Illustration of what the normal CSFF would have looked liked is juxtaposed next to the Encased CSFF:

Normal CSFF   **Encased CSFF**

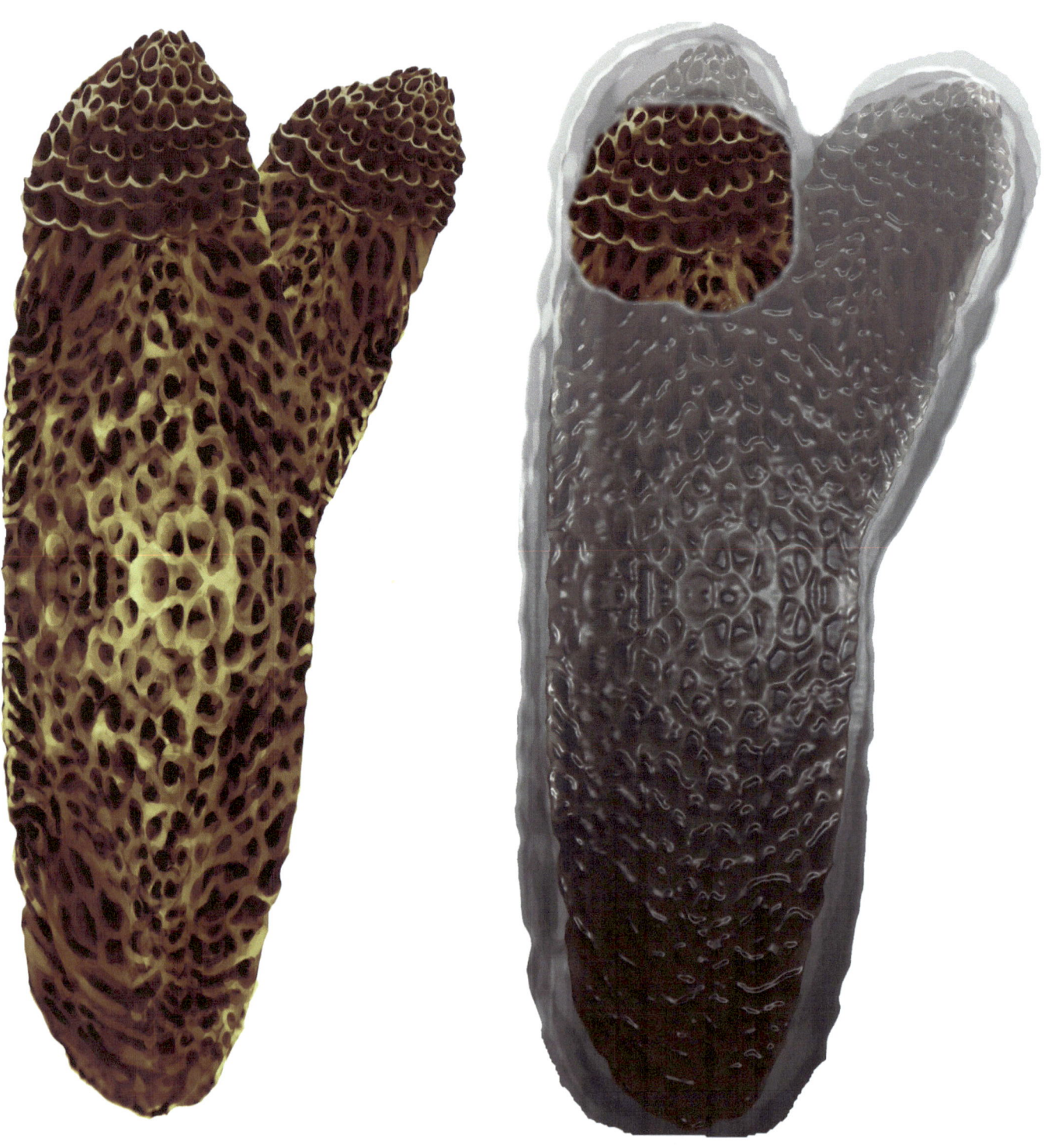

The actual fossil photos of the **Large Encased CSFF** are shown below:

Exterior View, Transverse Break in Middle

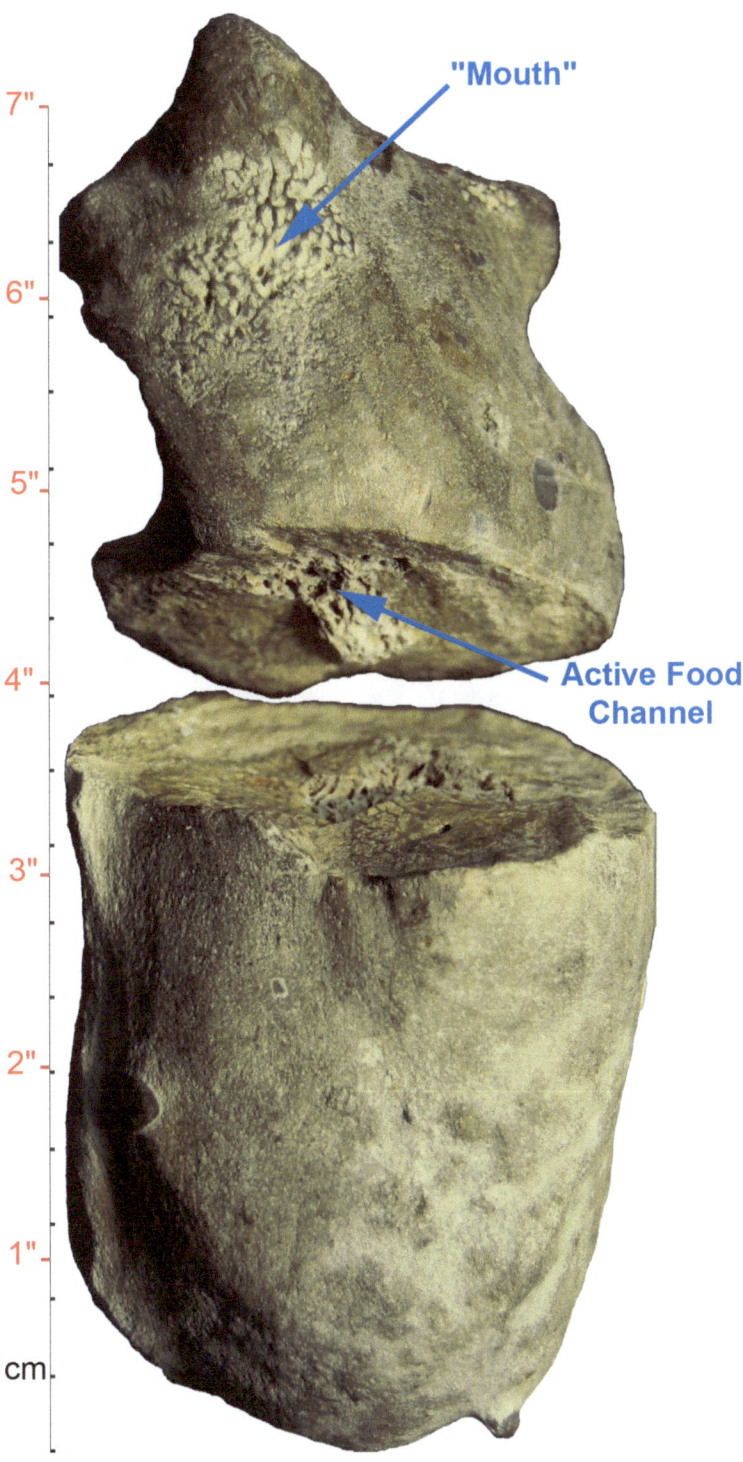

It appears the encasing colony prevented large parts of the underlying CSFF structure from obtaining food. The only exposed orifice to obtain food is at the top left (i.e. mouth). The only active food channel grouping is downstream of the mouth.

The photo below shows a close up of the "mouth" end to better show the orifice and active food trail.

Exterior View, Top Half, Matrix Removed

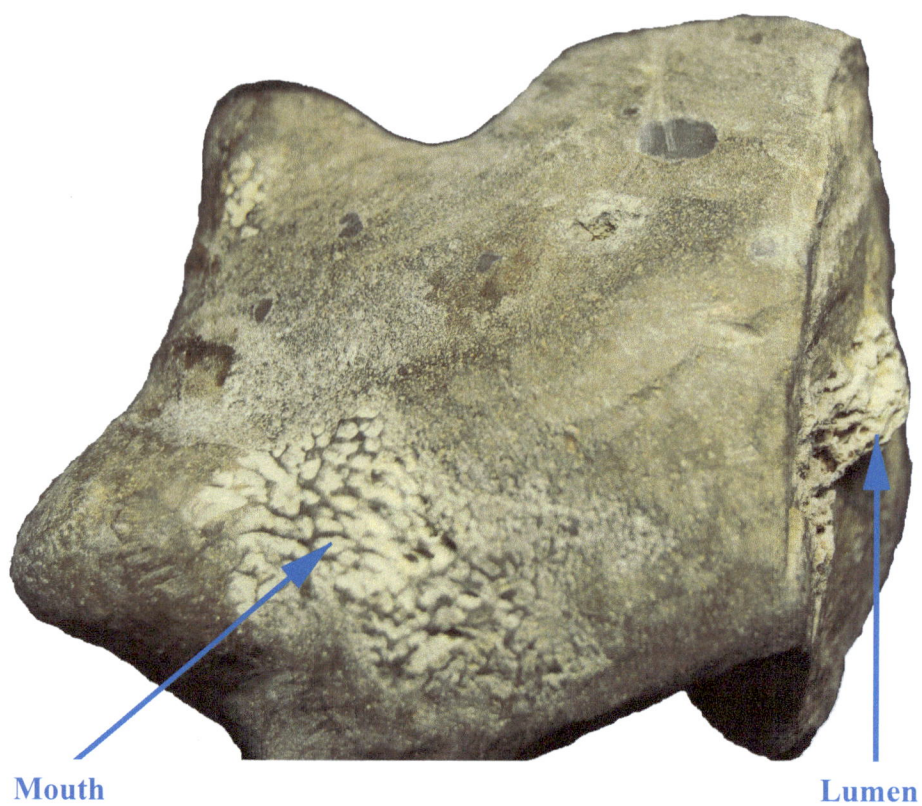

Mouth   Lumen

Transverse View, Looking Toward Mouth

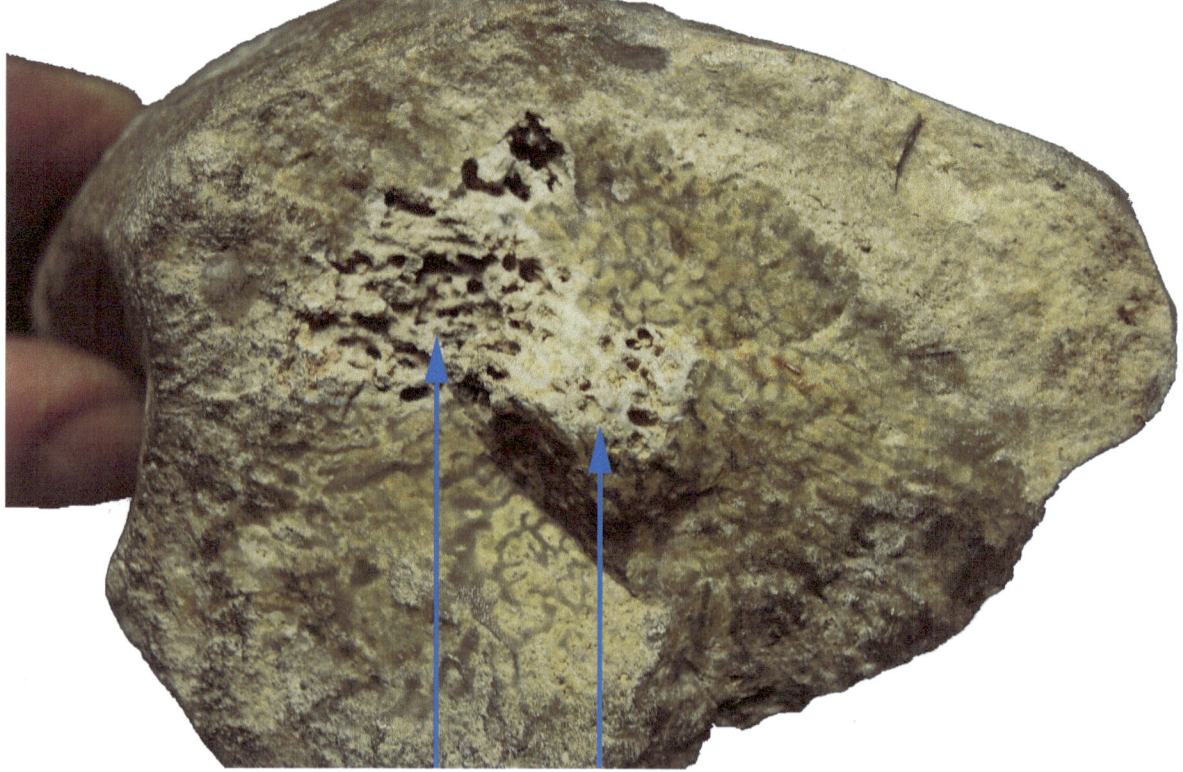

The transverse view shows the only active food channel group is downstream of the mouth. The implication is that the encasing cells caused the demise of the CSFF, rather than acting synergistically.

Another example of encasement is the **Encased Spherical CSFF**. It appears to be a **Spherical CSFF** that has become encased or surrounded by something, however by what is not clear.

Illustration

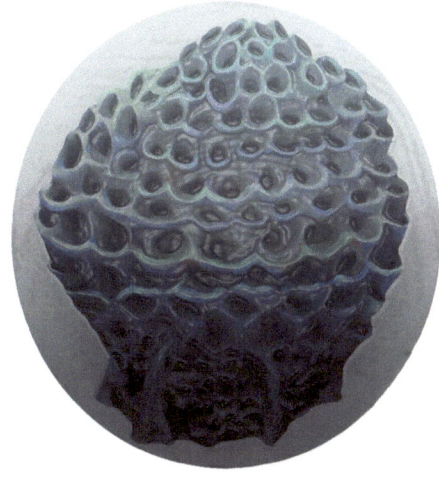

Sectioned Specimen Photo

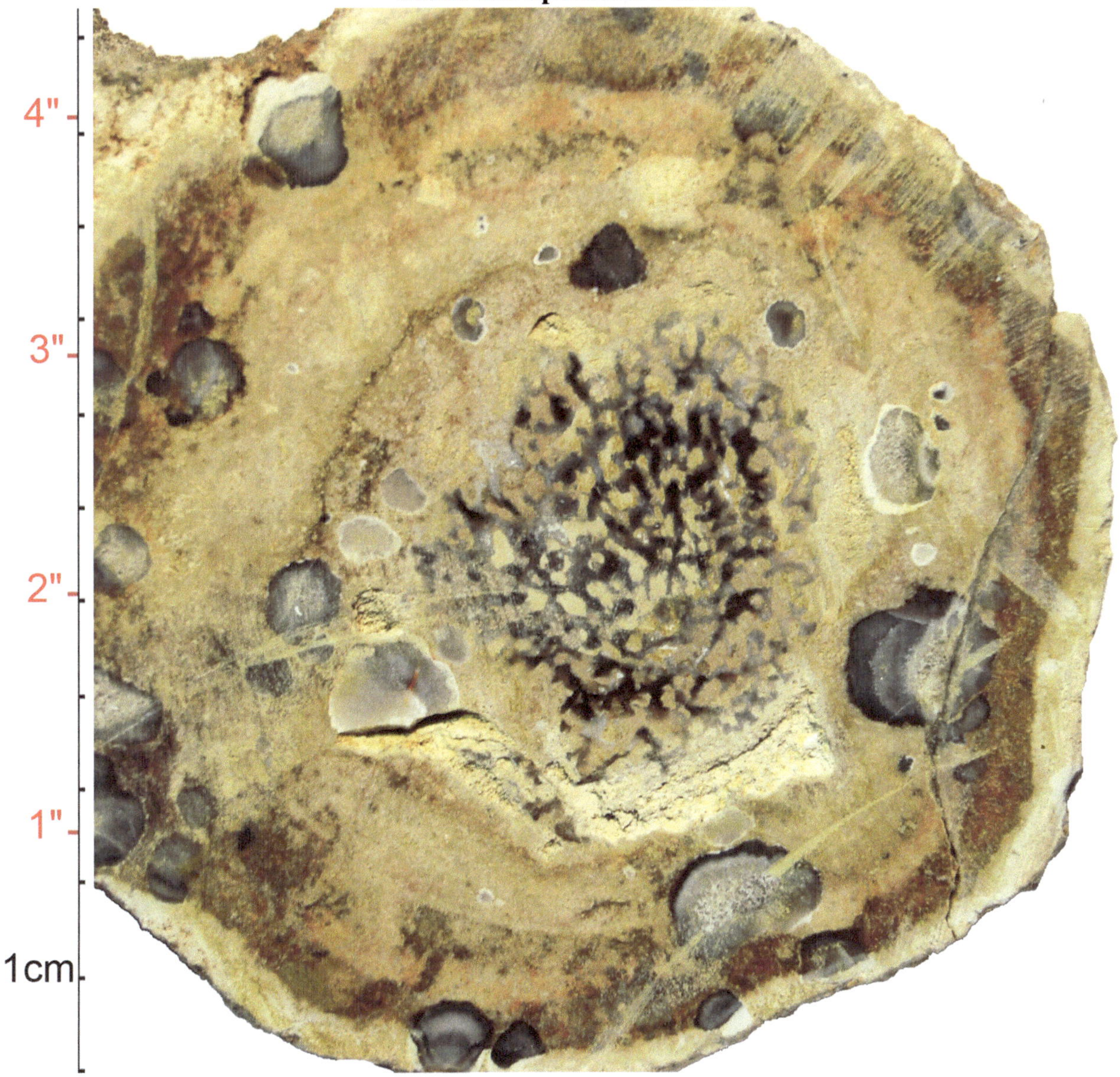

Another example is an encased **Protruding CSFF**, where the underlying encased organism or colony appears to have altered its construction plan to obtain food through tube like structures.

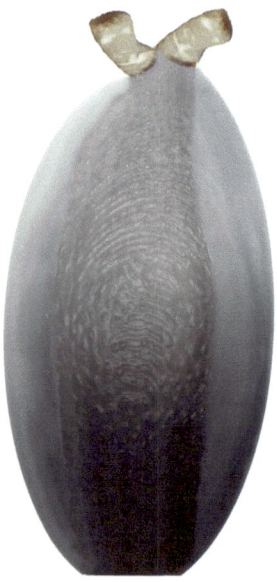

Internal View of Side A   **Internal View of Side B**

Zoom of Adapted Feeding Structures

It is not clear if both adapted structures above are for feeding or one is for feeding and the other for eliminating waste. It is also possible the tube may be a claw or "foot" structure for motility.

While the above three examples appear to be parasitic invasions from the outside, there is also evidence of parasitic invasion from the inside. The **Internal Soft Tissue CSFF** is an example. The CSFF may have served as an incubator for the internal life form, which may possibly be a forerunner of ferns or mosses.

The illustration below anticipates the demise of the CSFF, as the internal soft tissue structures are still bound inside the declining CSFF in the actual fossil.

Internal Soft Tissue CSFF Illustration

The photo below shows a side by side view of the external and internal structures. The exterior has a typical CSFF structure as previously seen. However, the sectioned interior view shows the soft tissue or leaf like structures dominating the interior of the life form.

Side A: Exterior View   **Side A: Interior View**

The illustration on the previous page anticipates the eventual demise of the CSFF exterior, and the "blooming" of the internal soft tissue from within.

A close up of the internal view shows soft tissue (leaf-like structures) being the dominant mass inside the CSFF structure:

Leaf-Like Structures Inside

Chapter Summary:

A portion of the hard skeleton CSFFs appear to have integrated with soft tissue.

The end result of all the integrations reviewed in this chapter is that the CSFFs were no longer able to perform the task of genetic reassortment.

It is not clear if the synergistic CSFF / soft tissue integrations went on to become viable life forms (e.g. sponges, bryozoans, plant life etc...) or simply went extinct.

While most of the life forms presented in Chapter 2 were morphologically similar to known life forms, a part of the viable CSFF progeny never made it into earth's play book of life. Just because a life form was viable was no guarantee it would withstand the test of time. These specimens are covered in the next chapter.

Chapter 5

Unicellular Giants and Indeterminate Life Forms

The simplest life forms resulting from the DNA reassortment process could be expected to be unicellular organisms.

Genetically reassorted cells with DNA that coded for less frequent cell division times, increased expression of growth promoting genes, or decreased expression of growth inhibiting genes, could result in giantism. Examples of both non flagellate and flagellate unicellular giants appear in these fossils.

An example of what appears to be a unicellular giant undergoing cell division is presented in the first example. A photo derived rendering of a **Dividing Unicellular Giant** is shown below.

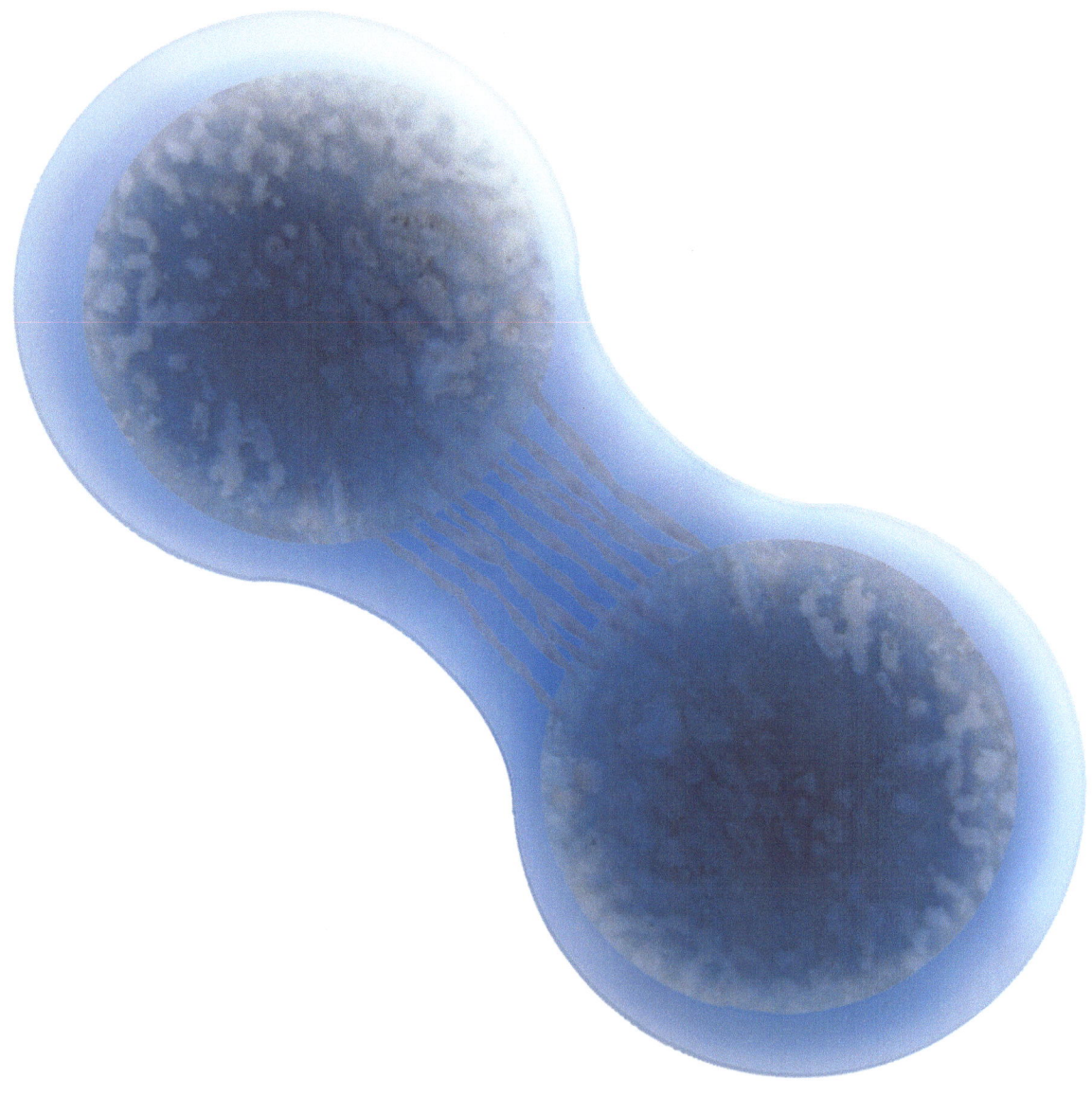

The photo of the **Dividing Unicellular Giant** below appears to show two cell divisions underway. The mother cell (labeled **1**) is at the very bottom, the younger progeny (labeled **2**) appears to be finishing the division process, and the older progeny (labeled **3**) appears to be in the final stage of pinching off (and possibly already starting a cell division of its own). The fast and furious pace of the cell divisions would indicate a nutrient rich ambient environment was present, which would favor survival of newly created life forms.

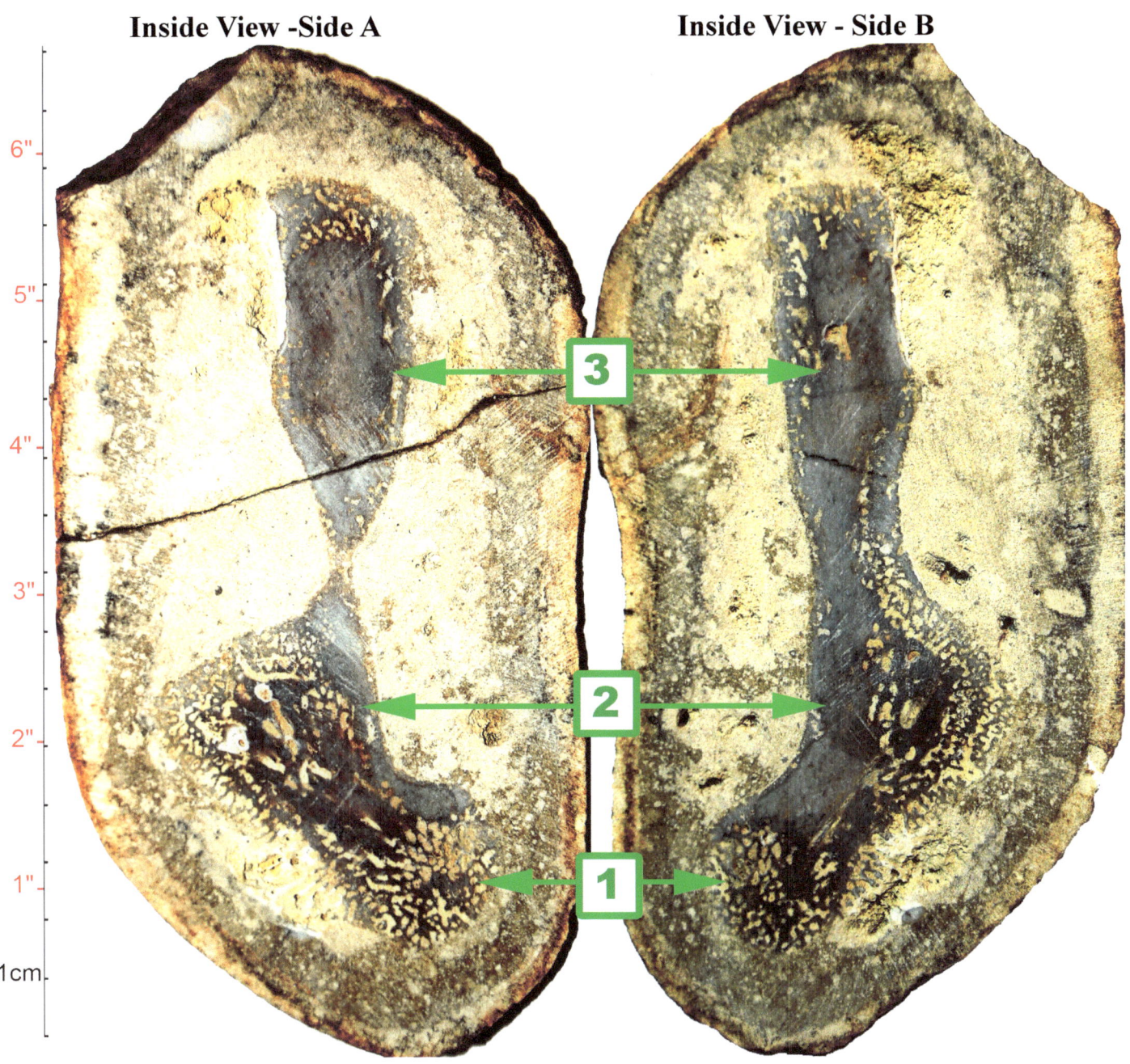

Possible vestiges of microtubules appear on Side A between **1** and **2** and the more clearly separated cells appear at **1** and **2** on Side B. The final stage of cell division is mediated by elongation (polymerization) of a set of microtubules from opposite poles, which push the two new cells apart. Microtubules are degraded (depolymerized) after they have served their purpose

An enlarged view of the apparent younger cell division, with possible microtubule vestiges, is shown below:

Lower Side A Zoom (in Matrix)   **Lower Side B Zoom (in Matrix)**

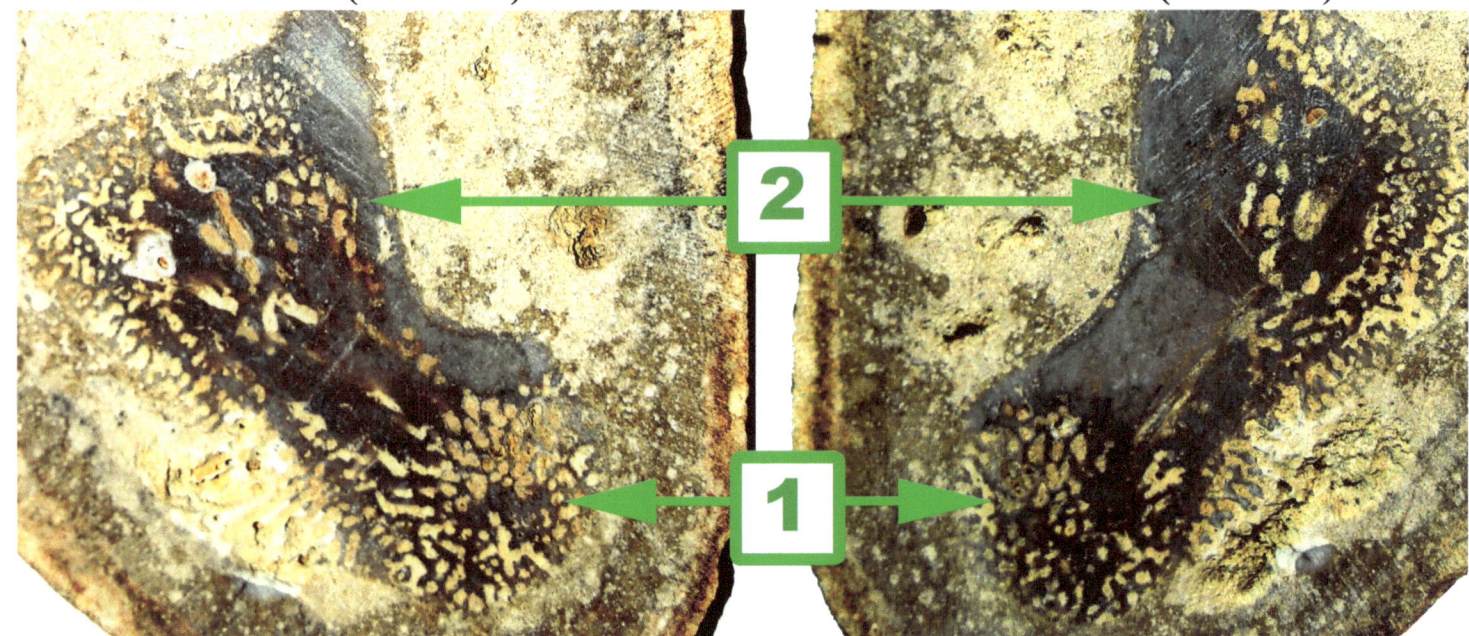

The two separated cells can be better seen on Side B at locations 1 and 2 and vestiges of possible microtubules appear on Side A between locations 1 and 2 . DNA expression / synthesis of the protein tubulin determines if microtubules elongate or contract. Above a critical concentration of tubulin in the surrounding fluid the microtubules elongate, pushing the two new cells apart. Below a critical level they disintegrate.

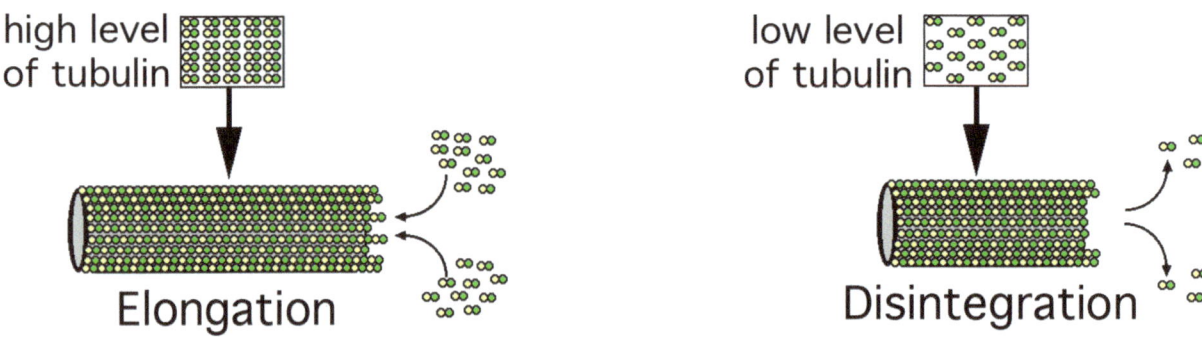

A public domain photo of a microscopic cell beginning its division *(by Roy van Heesbeen)* is shown below. In contrast, the unicellular giant in our photo above appears to be in the final stages of its cell division.

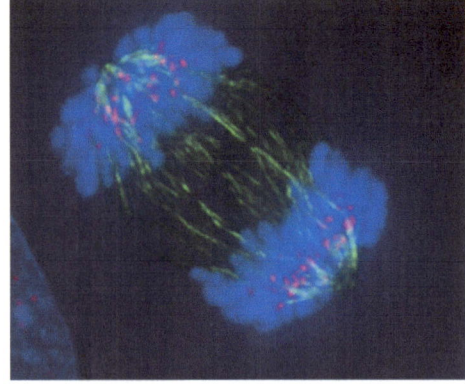

Another specimen, the **Flagellate Unicellular Giant**, is ~ 2" long, and has rear mount flagellate propulsion. The photo derived rendering was previously presented and is shown again below:

Flagellate Unicellular Giant

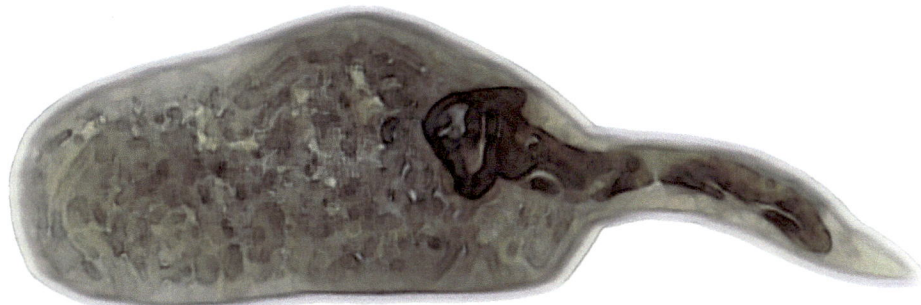

Enlarged photos of both sides of the actual specimen are provided below:

Enlarged View of Side A in Matrix

Enlarged View of Side B in Matrix

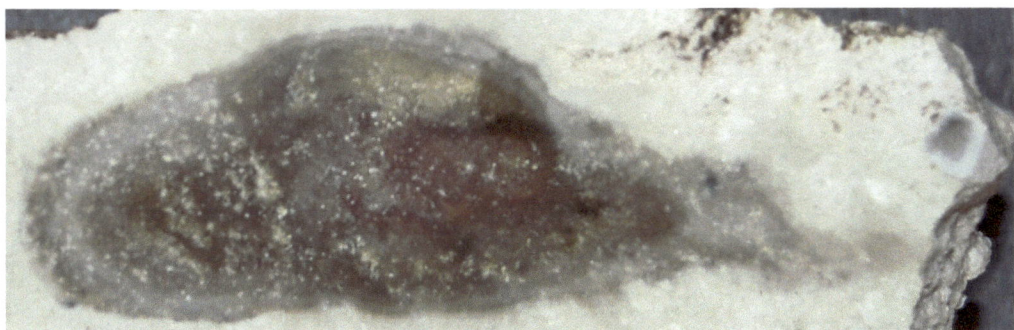

A photo of sectioned Side A with the matrix digitally excluded is shown below:

The propulsion system has been digitally highlighted below and it appear to be a giant version of the traditional rear mount flagellum / kinetoplast propulsion system.

A slight upgrade from unicellular giants are what author refers to as "The Reds".

The reds are characterized primarily by their globular appearance, without the more defined morphological features possessed by multicellular organisms. Their body shape more closely resembles unicellular giants.

Thumbnail illustrations of several Reds are shown below:

Red 1   Red 2   Red 3   Red 4   Red 5

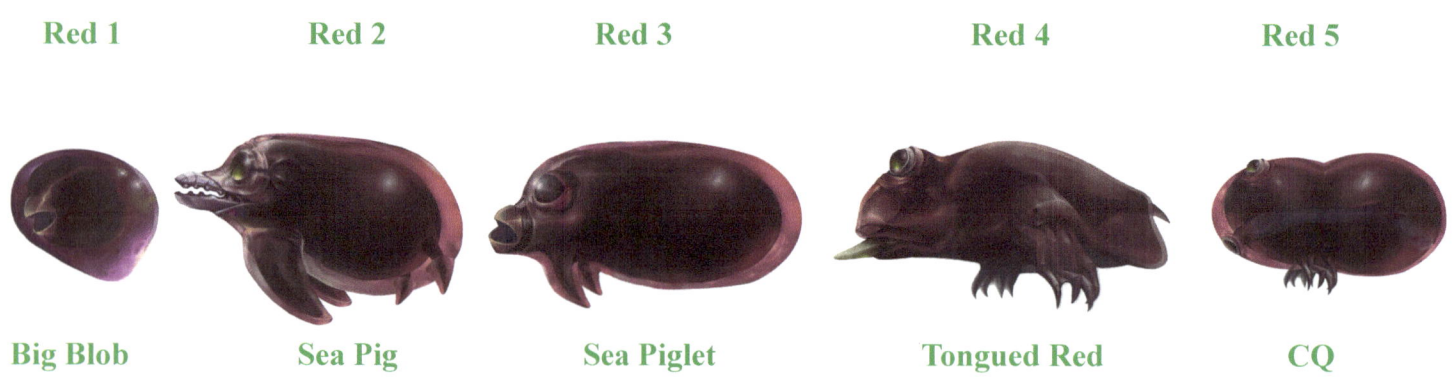

Big Blob   Sea Pig   Sea Piglet   Tongued Red   CQ

Red 1 or Big Blob is so named because it most closely resembles a giant single amorphous cell with no distinguishable morphological features other than its mouth.

The artist's rendering of **Big Blob** type organisms is presented below.

The actual fossil photos of Big Blob are shown below:

Red 1: Big Blob
3.25" (8 cm) long

Exterior View	**Interior View**

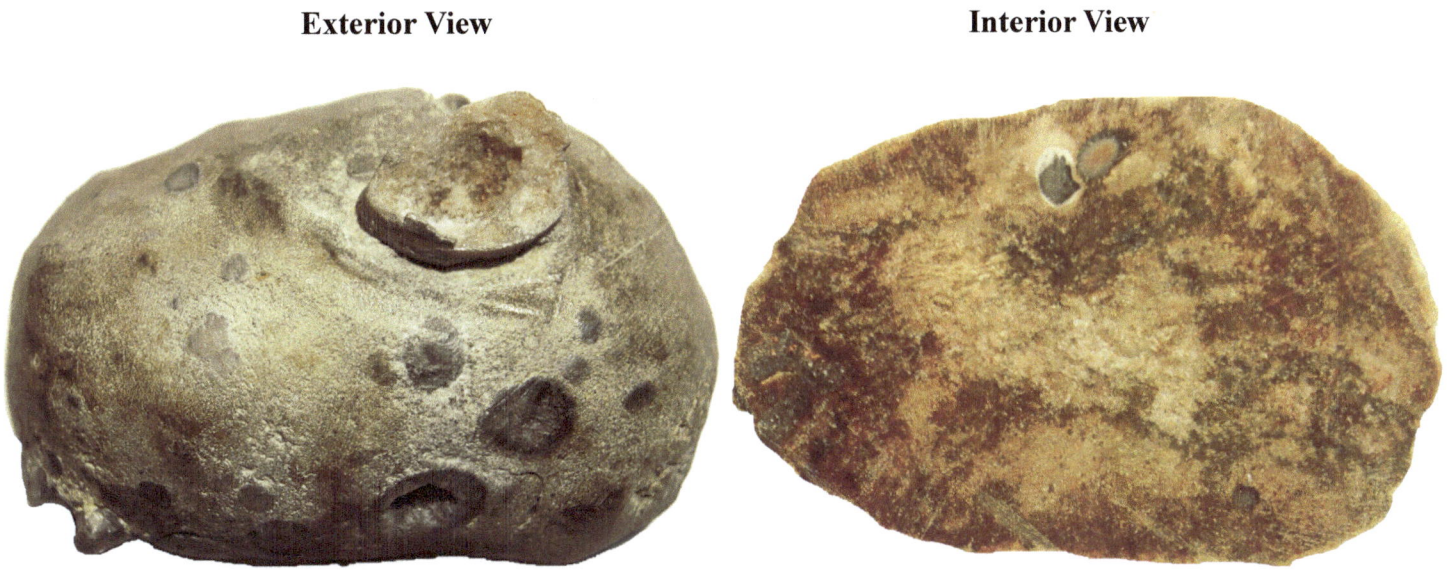

Unicellular blobs like this would likely have become "manna of the ocean" once jawed predators arose.

Red 2 or Sea Pig begins to have more interesting morphological features, the most notable being the more advanced motility system at the bottom of the organism. The artist's rendering is shown below:

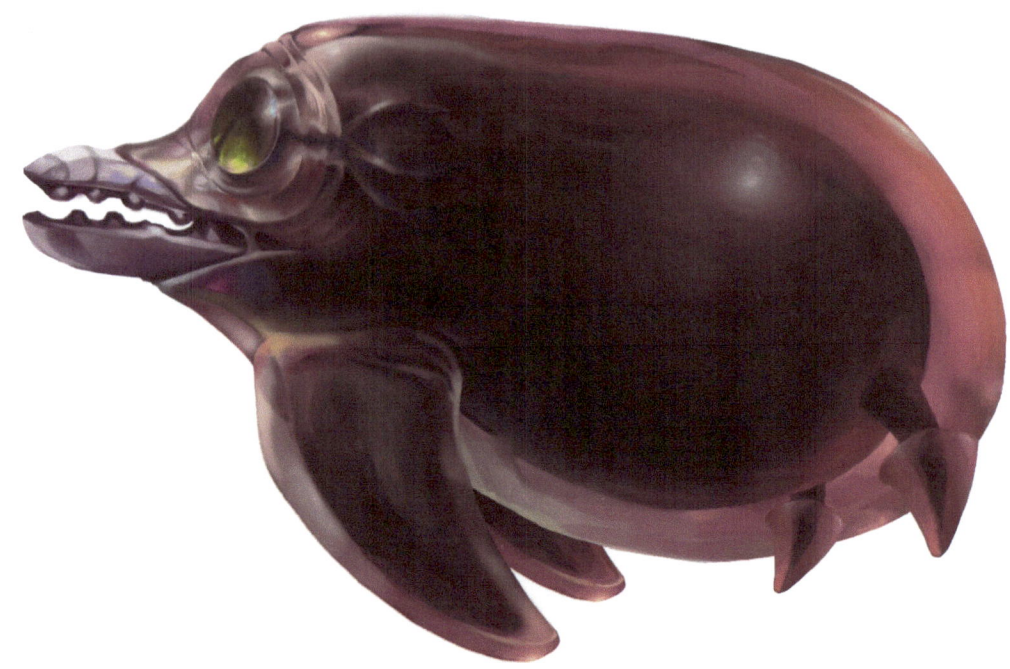

The actual sectioned fossil photo of **Sea Pig** is shown below. The specimen is 2.5" (6.5 cm) long.

Red 2: Sea Pig
Interior View of Side A, Matrix Excluded

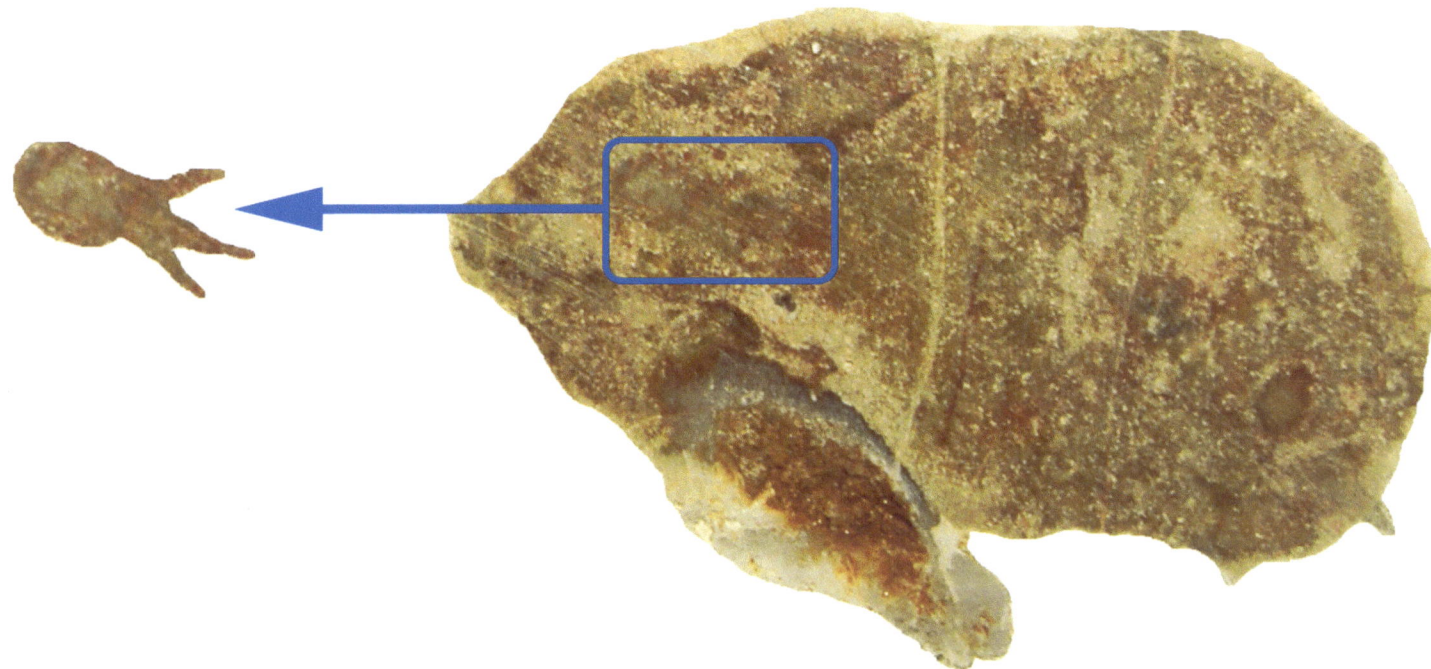

A more subtle, but more significant feature, resides where the eye would be expected to be. It appears to be a circular organelle with rearward protruding appendages, possibly being a truncated forerunner to the optic neuron reviewed in **Whaleen**. The optic structure is isolated and enlarged for clarity above.

Red 3 or Sea Piglet is morphologically similar to Red 2, but is smaller and has less advanced front flippers / pseudopods. It is not clear if is a young version of Red 2 or another species.

An illustration of Sea Piglet:

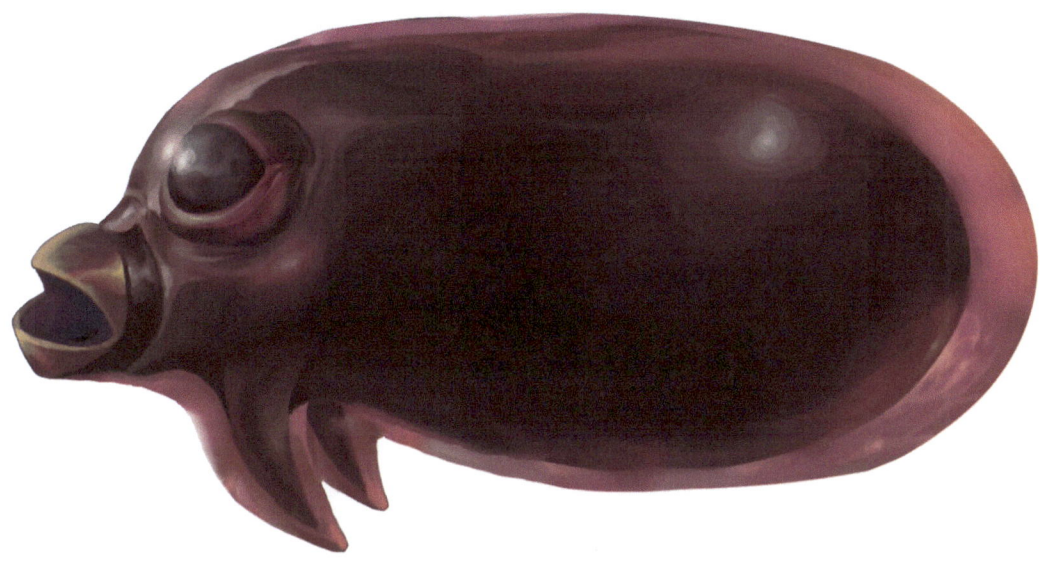

The actual fossil photo of Sea Piglet:

Red 3: Sea Piglet
Sectioned Side A, Matrix Excluded
1.7" (4.3 cm)

Red 4 or Tongued Red appears to have a cycloptic eye, a tongue or proboscis, and claws.

Both halves of the actual sectioned specimen are shown below, with Side A on top, and Side B on bottom. **Tongued Red** is 2.25 " long. The head has has been oriented to the left in both photos.

Sectioned Specimen Photos

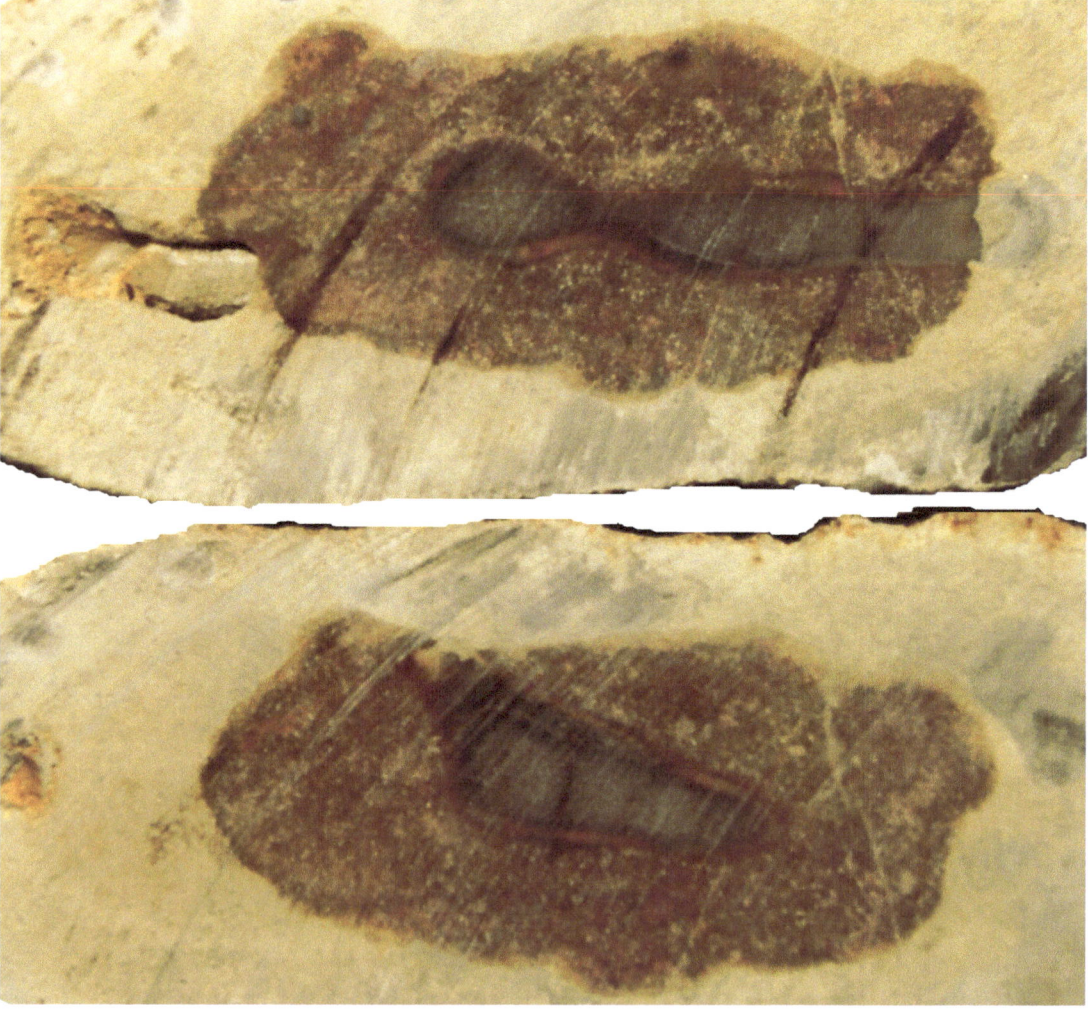

Although the protrusion is assumed to be a tongue or proboscis, it may also be a worm or other piece of food Red 4 was eating when it perished or it may also be a worm or parasite eating Red 4 after its demise.

Tongued Red, with the matrix digitally excluded is shown below. Claw like structures can be seen at the bottom center of the life form.

Sectioned Side A, Matrix Excluded

Sectioned Side B, Matrix Excluded

Tongued Red also appears to have something segmented in its central cavity, but because it appears to be pinching off in Side A, it may likely be progeny. Alternatively, it may just be the remains of whatever the tongued red ate.

Red 5 or Central Question (CQ) appears to have something long and segmented in its central food lumen.

Illustration of **Central Question**:

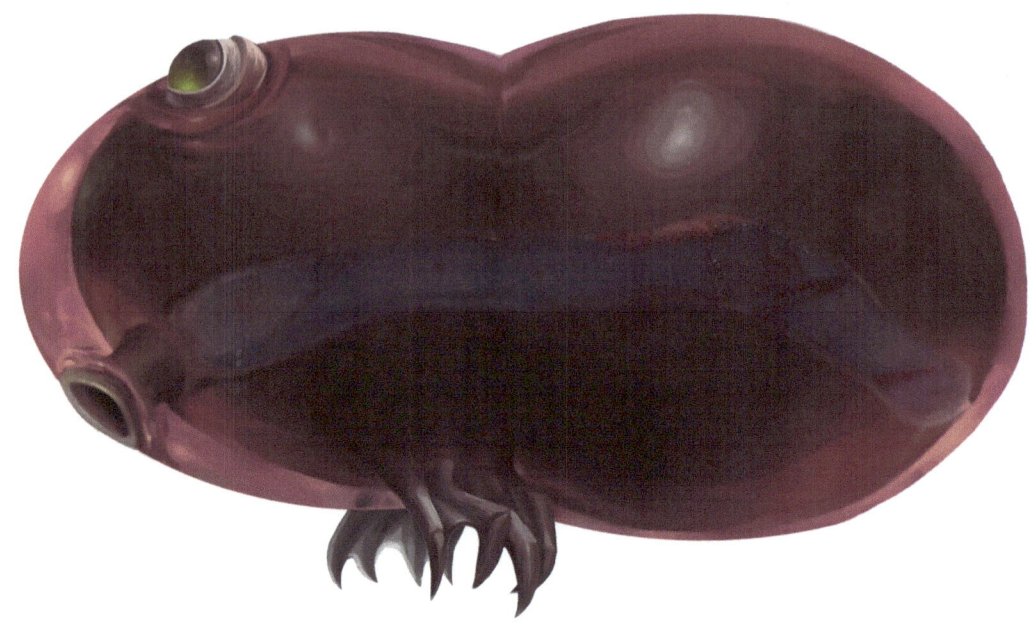

Photo of **Central Question**:

Red 5: Central Question
Sectioned Side A, Matrix Excluded
2.7" (6.9 cm)

An isolated zoom of the segmented structure or objects inside CQ are shown below:

The segments do not appear to be progeny, as there is no pinching off visible.

Another possibility is that this structure may be a forerunner of vertebrae.

However, the most likely explanation may be simply that the segments are contents of what Central Question (CQ) ate. The limited motility of CQ (small claws appear to be more for attachment than motility) and absence of jaws may have doomed it to feed on available plant life, stationary shelled life forms, or scavenge debris off the ocean floor.

Pieces of an organism that could match the contents of the lumen is shown in the next specimen called **The Y**. An illustration of **The Y** is shown below:

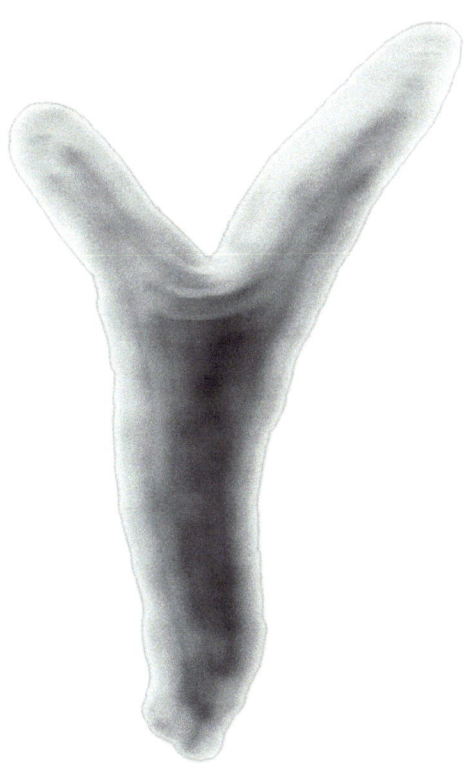

It is not clear if **The Y** is a plant or shelled animal. It may be a type of bryozoan. The inside of the organism is dark and surrounded by a white outer shell. The specimen itself has been removed and is shown below. It is only around a quarter inch thick.

The Y
Interior View of Sectioned Halves

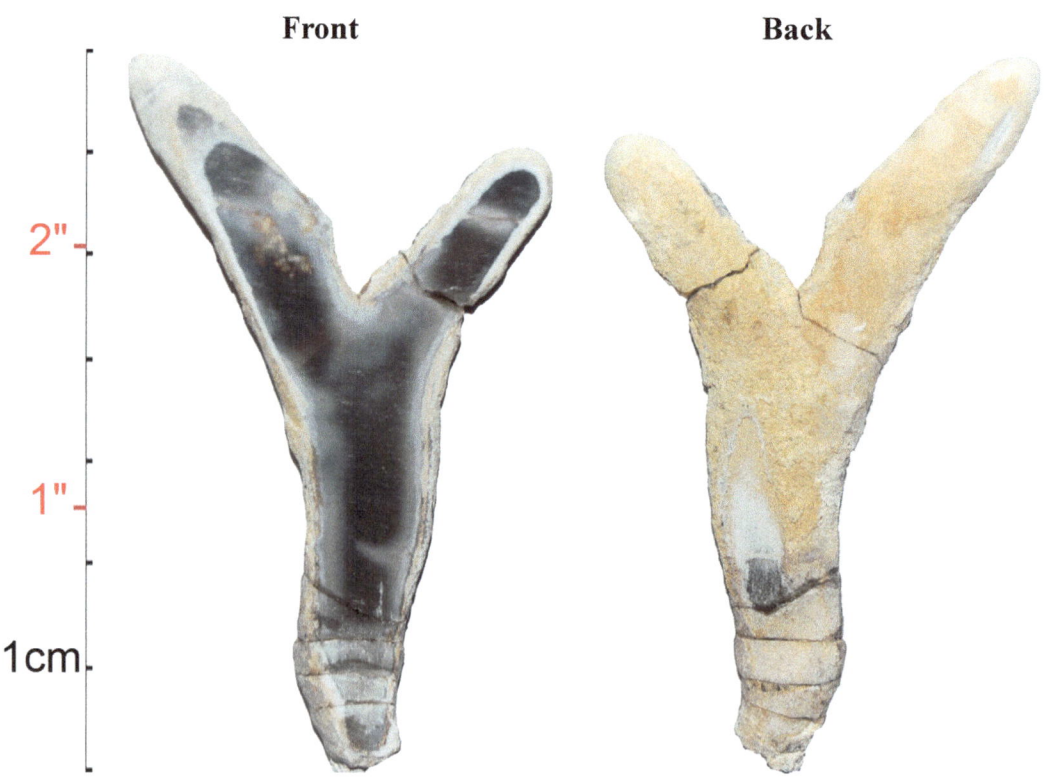

Specimen Removed From Matrix

Front   Back

2" -

1" -

1cm.

A zoom of the front side of **The Y** is shown below. The pure white shell can be seen surrounding the organism (especially visible on the right top branch of the Y).

Sectioned Specimen

The next specimen cannot be interpreted with certainty. Protist digest food in a central vacuole, indicating this may be a giant teardrop shaped protist. An image of a life form resembling a young **Sea Snake** appears at the bottom, suggesting the surrounding structure may instead be an egg. Alternatively, the specimen may also be a large protist that recently ate and is digesting a small sea snake like creature.

Sea Snake in Egg or Teardrop Shaped Protist
Sectioned View, in Matrix

The isolated **Sea Snake** life form at the bottom of the specimen is shown below:

The artist's rendering of what Sea Snake may have looked like is shown below:

Cells with redundant and / or incompatible DNA could be expected from the grab bag process. Most would not be viable but a rare few could go on to maturity. **Double Fish** appears to be one such example.

Photo of Exterior View of Side A (~ 8" long):

Photo of Interior View of Side A:

The external view may appear normal however the internal views tell a different story. Sectioned Side B has the pincer-like outer mouth closed, but a zoom reveals a second redundant internal "jawed" fish (gray):

Side B - Internal View
Dual Jawless and Jawed Mouth Structures

Pincered Outer Mouth Closed

Zoom of Internal Jawed Mouth

The gray jawed fish runs the entire length of the surrounding beige fish, as seen in sectioned Side B below.

The gray jawed fish on the left appears to have a separate U shaped lumen, as opposed to the external beige "pincer" mouthed fish on the right side, which has a central linear lumen. Had the saw blade been a quarter inch to the right, the two fish would have been cleanly separated.

Sectioned Side B - Transverse View

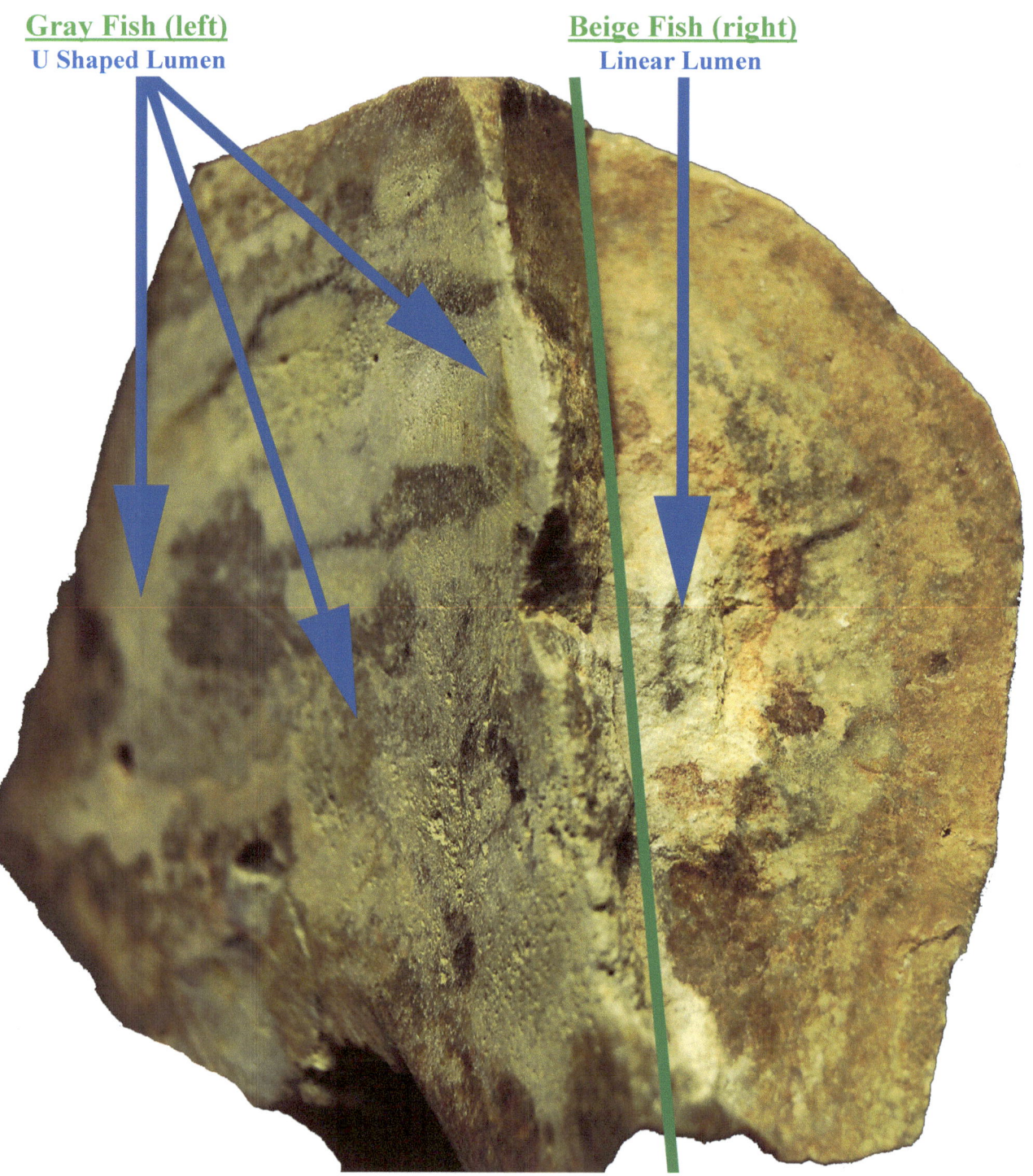

Life forms with redundant and / or different structures, such as in Double Fish, that survived to maturity, would be rare, but not unexpected. The lack of boned structures likely allowed the more amorphous body structure to accommodate both life forms.

Aqua Duck is missing the rearmost part of the tail and front most part of the mouth. There appears to be a large flipper that starts at the mid body and extends all the way to the rear of the animal.

Aqua Duck

Side A -Exterior View

Tail   Flipper / Wing Forerunner   Head

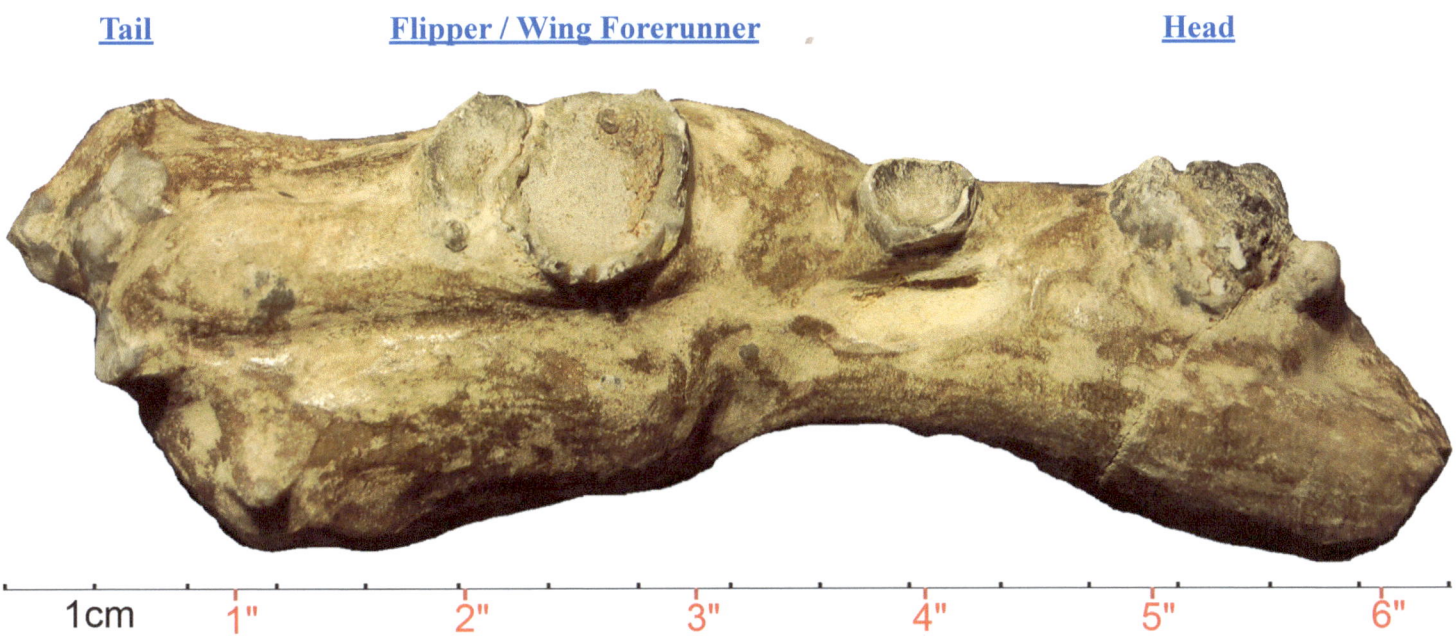

1cm   1"   2"   3"   4"   5"   6"

The artist's rendering of what Aqua Duck may have looked like is shown below:

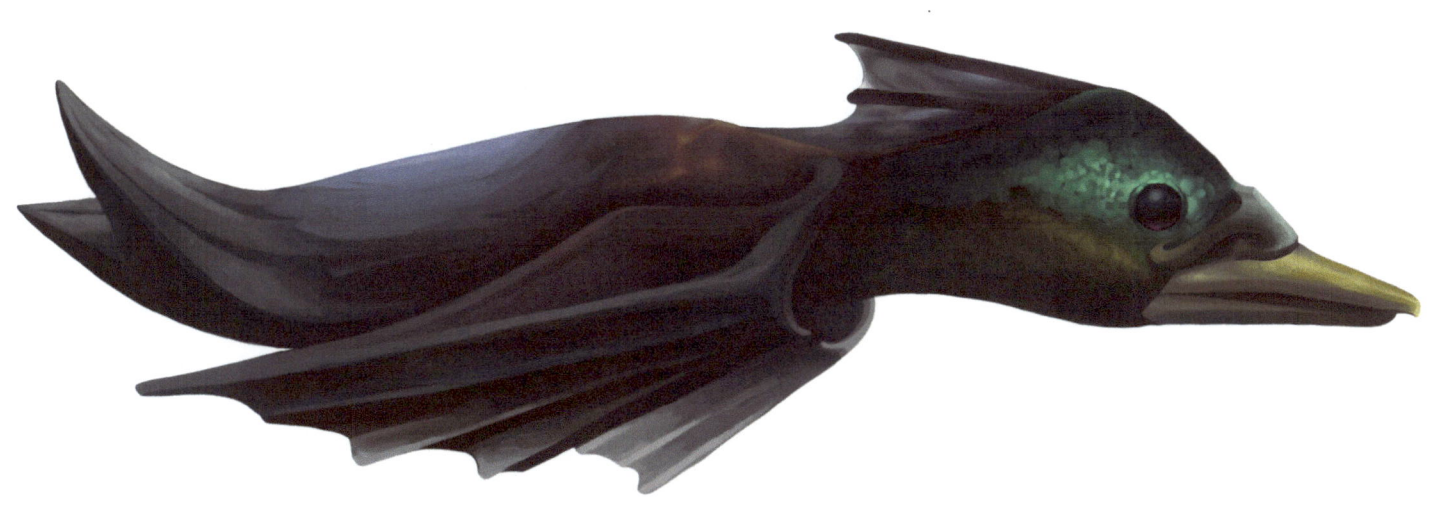

Sea Horse has a head and body shape similar to a sea horse, but the body is much larger than current day sea horses. The artist rendering of the specimen is shown below:

The actual specimen photos are shown for reference:

Exterior View - Side A

The interior view of **Sea Horse** reveals a typical central food lumen.

Interior View - Side A

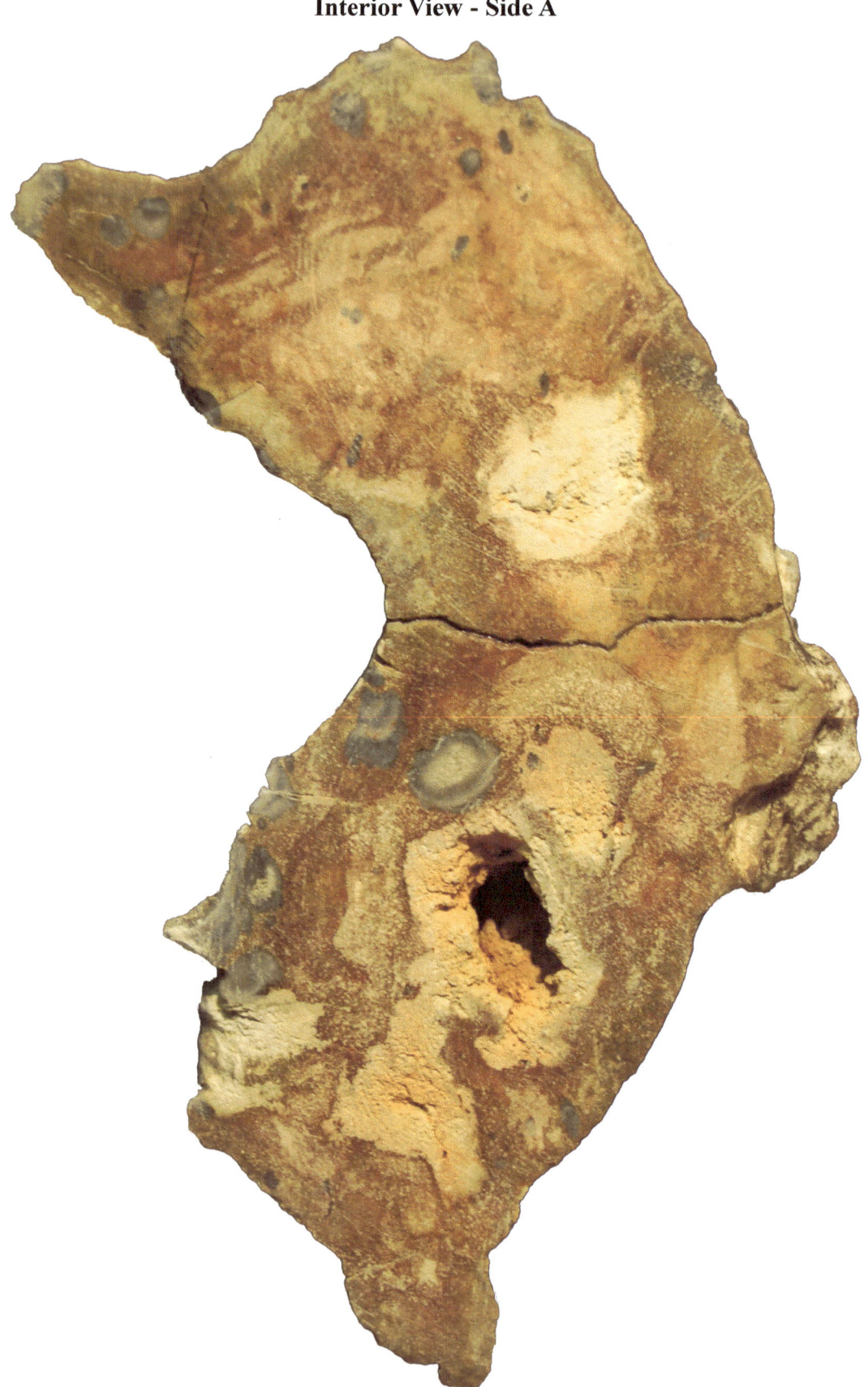

94

Another indeterminate life form presented is called **Eel-a-Phant**. The artist's rendering of   **Eel-a-Phant** is shown below. The actual specimen is ~ 5" long.

Illustration of **Eel-a-Phant**:

Photos of **Eel-a-Phant**:

Side A - Exterior View   **Side A - Interior View**

The trunk or finger like structure (on left) may have been used to gather food and pull the organism along.

The organism inside appears similar to an eel.

The last specimen appears to be a Jawless Fish. Jawless fish also appeared in the Cambrian period, but were believed to have gone extinct during the Devonian period.

Sectioned Specimen, In Matrix

Artist's rendering of Jawless Fish:

Chapter Summary: The biodiversity inherent in the fossils presented in this chapter further supports the random (grab bag) genetic reassortment process proposed by the author. Final thoughts, conclusions, and implications are presented in the next and final chapter.

Chapter 6: Summary and Conclusions

The fossils, when combined with molecular biology and fluid dynamics, reveal a likely process that underlies the origin of biodiverse, multicellular life.

Combined with what we know about earth, we can conclude the following:

- **Prokaryotic cells appear to have been the seeds of life**

- **They (cyanobacteria) transformed earth into a planet with an oxygen based atmosphere**

- **They provided DNA and mitochondria for eukaryotic cells**

- **A class of eukaryotic cells (CSFFs) are capable of "grab bag" genetic reassortment**
  - **As an unintended consequence of how they feed (obtain intracellular nutrients)**

- **This "grab bag" reassortment process generates biodiverse multicellular life**
  - **By a process separate and distinct from evolution**

- **The CSFFs may also have been the forerunners of bone building cells in animals**
  - **Bone / Soft Tissue integration is crucial for higher multicellular life**

Although the DNA reassortment process that generates biodiverse life is separate and distinct from the evolutionary process, it effectively serves as the starting point for the subsequent evolutionary process. This DNA reassortment driven "Quantum Speciation" process fills the gaps that cannot be explained by evolution.

The absence of boned structures, but morphological similarity of the specimens to life forms in existence today, suggests these life forms may be an intermediate step between unicellular life and vertebrate multicellular life.

The fossils provide evidence that the CSFF driven genetic reassortment process repopulated the earth after one known extinction event. It is not clear how massive an extinction event needs to be before the CSFFs can once again establish themselves. It is not even clear if any of the CSFF unicellular eukaryotes are still alive today.

The next step from an experimental perspective would be to recreate the genetic reassortment structures observed, recreate the environmental conditions present in the Permian and Cambrian periods, and see if the predicted genetic reassortment can be recreated in the lab. The further back in time one goes, the closer the moon was to earth, and the larger the tides were. The moon's proximity 3.8 billion years ago resulted in 1,000 foot tides. It is not clear if today's subdued fluid dynamics would be adequate to drive a biodiversity explosion event, even if the CSFFs were to reestablish themselves after a mass extinction event.

The most important question however goes back to our prokaryotic ancestors, which today also reside in or on our bodies at greater cell counts than our own. We basically serve to feed them, and one day may serve to also distribute them to other planets.

Prokaryotic cells first appeared 3.8 billion years ago, only 0.1 billion years after the end of the asteroid impacts. This makes the theory of bacteria being brought to earth by asteroid impacts a good possibility. That in turn implies that every time an asteroid hits the earth, particularly in the ocean, an entirely new collection of viable DNA may be introduced into the genetic reassortment process. Accordingly, many of the post extinction progeny may not resemble anything that lived prior to the extinction event.

Eukaryotic evolution over 1.5 billion years has not been able to result in a cell that has the molecular biology sophistication of prokaryotic cell (i.e. genomic efficiency, triple layer wall structure, its own mitochondria).

The molecular biology implication is that eukaryotic life on earth hails from prokaryotic origins.

Which leaves the final planetary origins of life question: Where did our ancestral bacteria come from and how were they made?

Acknowledgments

Author would like to thank the following for their generosity in time and contribution:

Dr. James T. Sprinkle, Ph.D., University of Texas at Austin, whose unique knowledge and insights, pointed to a potential nexus between the Cambrian quantum speciation event and the Permian quantum speciation event.

Dr. Justin John Zamoyski, D.O., for his review of the molecular biology, neurology, the bone microenvironment and contributions related to fluid dynamics which were applicable to the mechanistic operation of the genetic reassortment structures.

Anne L. Neeter, Esq. , for her insights and advice on both copyright laws and e-publishing markets.

Dr. Bijan Gillani Ph.D., California State University - East Bay, for his review, insights, and advice with respect to this book.

U.S. Geological Survey (USGS), for its publicly available Time and Terrain maps, which make invaluable scientific information available to the public.

Centers for Disease Control and Prevention, **Public Health Image Library**, US Department of Health and Human Services, for the public domain photo by **Dr. Myron G. Schultz** of the flagellate protozoa *trypanosoma*.

Roy van Heesbeen for his public domain photo Anaphase_IF from Wikimedia Commons

Judy Gillani, for her editorial review of the book

Dr. Richard Aplin, Ph.D., whose actions affecting me more than 30 years ago set in motion a series of events that have resulted in many beneficial outcomes, including the publication of this book.

About the Author

Mark Zamoyski received his BS in 1977 and MBA in 1978, both from Cornell University (Ag. & Life Sci. and JGSM, respectively). His scientific curiosity in the history of life on earth started several decades ago from fossils prevalent on one of the family properties in Arizona. In 2011, Mark set up specialized equipment to process the fossil collection in his spare time and started evaluating the collection to understand the story it told. The result is this book.

In his professional career, Mark has been awarded 15 US patents, most dealing with the molecular biology of disease conditions and novel treatment methods based thereon.

About the Illustrator

Nicholas J. Lee is a illustrator of ideas and inventions. He lives a peaceful life amidst books and long journeys.

Appendix A
Bacterial Genomic Efficiency

Humans have 46 chromosomes and 3 billion base pairs (bp) which code for the production of an estimated 21,787 proteins *(DOE, Human Genome Project Information, October 2004 findings)*.

Bacteria typically have a single circular chromosome, however some have 2 circular chromosomes and some have combinations of a circular and linear chromosome. Many bacteria also have plasmids, which are separate strands of DNA in the cytoplasm.

Genomic efficiency of bacteria is based on samples from data available on the UniProt (Universal Protein Resource) database as of August 2011:

	DNA Base Pairs (bp)	Proteins Known	**Proteins Per Million bp**	% bp Chromosomal
Cyanobacteria				
CYAA5	5,460.377	5,239	959	98%
CYAP8	4,787,694	4,335	905	98%
CYAP0	4,803,347	4,045	917	97%
CYAP4	5,786,110	5,232	904	93%
CYAP7	6,554,169	5,656	863	91%
CYAP2	7,841,948	6,559	836	78%
Average	5,872,274	5,238	**898**	
Rickettsia				
RICPR	1,111,523	834	750	100%
RICRO	1,268,188	1,385	1,092	100%
RICRS	1,257,710	1,345	1,069	100%
RICPP	1,111,612	952	856	100%
RICTV	1,111,496	837	753	100%
RICAH	1,231,060	1,257	1,021	100%
RICB8	1,528,980	1,443	944	100%
RICBR	1,522,076	1,400	920	100%
RICCK	1,159,772	1,091	941	100%
RICCN	1,268,755	1,372	1,081	100%
RICAE	1,290,917	1,041	806	99%
RICPU	1,314,898	927	705	98%
RICM5	1,376,184	969	704	99%
RICFE	1,587,240	1,428	900	94%
Average	1,295,744	1,163	**896**	

Burkholderia strains include plant growth promoting strains (species codes BURPP, BURP8), plant disease causing strains (species code BURRH), and human disease causing strains (BURMA). Burkholderia typically have 2 circular chromosomes, and some have fairly large plasmids (BURP8 has a 1,904,893 bp and 595,108 bp plasmid).

	DNA Base Pairs (bp)	Proteins Known	**Proteins Per Million bp**	% bp Chromosomal
Burkholderia				
BURTA	6,723,972	5,561	827	100%
BURMA	5,835,527	4,797	822	100%
BRUMS	5,232,401	4,981	952	100%
BURP6	7,825,894	7,102	908	100%
BURRH	3,750,138	3,859	1,029	73%
BURPP	8,214,658	7,197	876	99%
BURP8	8,676,562	7,461	860	71%
Average	6,608,450	5,851	**896**	

Archaea are the oldest known prokaryotic cells, once classed with bacteria and named archaebacteria but now classified as a separate domain called archaea, has an even better genomic efficiency rate. Several different species were arbitrarily selected from the UniProt database. They all turned out to have a single circular chromosome and only one had a small 8,285 bp plasmid.

	DNA Base Pairs (bp)	Proteins Known	**Proteins Per Million bp**	% bp Chromosomal
Archaea				
ARCFU	2,178,400	2,398	1,101	100%
ARCPA	1,560,622	1,818	1,165	100%
ARCVS	1,901,943	2,065	1,086	100%
FERPA	2,196,266	2,463	1,121	100%
METM5	1,789,046	1,821	1,018	99.5%
Average	1,925,255	2,113	**1,098**	

References Cited

Alberts, B., Bray, D., Lewis, J., Raff, M., Roberts, K., Watson, J., Molecular Biology of the Cell, Third Edition, Garland Publishing, 1994

Blakey, R. and Ranney, W., Ancient Landscapes of the Colorado Plateau, Grand Canyon Association, p. 16, 145, 2008

Boardman, R. S., Cheetham, A. H., Rowell, A.J., Fossil Invertebrates, Blackwell Scientific Publications, 1987

Gibbs, M.E., Hutchinson, D., Hertz, L.: "Astrocytic involvement in learning and memory consolidation", Neruosci. Biobehav. Rev. 32, 927 - 944, 2008

Li and Dickie - K.W.K. Li and P.M. Dickie, Distribution and Abundance of Bacteria in the Ocean, Department of Fisheries and Oceans, Canada, 1996

Pereira, Alfredo Jr. and Fabio Augusto Furlan, "On the role of synchrony for neuron-astrocyte interactions and perceptual conscious processing", J Biol Phys, 35: 465-480, 2009

Sahney, S and Benton M. J., "Recovery from the most profound mass extinction of all time", Proceedings of the Royal Society B, 275, 759 - 765 , 2008

ScienceDaily,"Coral Reefs Made Rapid Comeback After the Greatest Mass Extinction of All Time", Oct. 5, 2011

www.ingramcontent.com/pod-product-compliance
Lightning Source LLC
Chambersburg PA
CBHW050727180526

45159CB00003B/1147